SOLID AND HAZARDOUS WASTE MANAGEMENT

SOLID AND HAZARDOUS WASTE MANAGEMENT

Science and Engineering

M.N. Rao

Razia Sultana

Sri Harsha Kota

BSP BS Publications
A Unit of BSP Books Pvt. Ltd., India.

ELSEVIER

AMSTERDAM • BOSTON • HEIDELBERG • LONDON
NEW YORK • OXFORD • PARIS • SAN DIEGO
SAN FRANCISCO • SINGAPORE • SYDNEY • TOKYO

Butterworth-Heinemann is an imprint of Elsevier

BH

Butterworth-Heinemann is an imprint of Elsevier
The Boulevard, Langford Lane, Kidlington, Oxford OX5 1GB, United Kingdom
50 Hampshire Street, 5th Floor, Cambridge, MA 02139, United States

British Library Cataloguing-in-Publication Data
A catalogue record for this book is available from the British Library

Library of Congress Cataloging-in-Publication Data
A catalog record for this book is available from the Library of Congress

ISBN: 978-0-12-809734-2

For Information on all Butterworth-Heinemann publications visit our website at https://www.elsevier.com

Publisher: Joe Hayton
Acquisition Editor: Ken McCombs
Editorial Project Manager: Peter Jardim
Production Project Manager: Kiruthika Govindaraju
Cover Designer: Mark Rogers

Typeset by MPS Limited, Chennai, India

CONTENTS

ABOUT THE AUTHOR

DR. M. NARAYANA RAO

Technical Consultant, Bangalore (India) Emeritus Professor, J.N. Technological University (JNTU), Hyderabad (India)

Dr. M. Narayana Rao is presently a technical consultant and an emeritus professor at JNT University, Hyderabad. He obtained his PhD in Environmental Engineering and did postdoctoral work in New Zealand. He taught at IIT Roorkee and NIT Durgapur for 16 years and later moved into Industrial R&D at Bangalore. He then joined Govt. of India and served as professor and later director of National Institute for Technical Teachers Training and Research, Chennai for about 10 years. Dr. Rao published 270 Technical papers in National and International Journals, 25 manuals, 5 textbooks, and 15 video lessons. He worked as an expert member on several Technical Committees of Pollution Control Boards and Ministries of Govt of India. and Govts of Andhra Pradesh, Tamil Nadu, West Bengal, and Karnataka. He is a consultant to UNEP, WHO, World Bank, Asian Development Bank, DANIDA, CIDA, Colombo Plan Staff College, Manila, and more than 60 industries in India and abroad. He is a member of academic councils of 9 universities and visiting professor to 10 Academic Institutions and Engineering Colleges. Dr. Rao guided about 250 Masters and PhD projects. He is a recipient of 25 awards including 4 gold medals. He chaired 40 International Conferences and presented Technical papers in 54 countries around the world. His areas of interest are water supply, wastewater treatment, air pollution control, waste management, climate change, EIA and Environmental Education. Email: narayanaraom@yahoo.com.

DR. RAZIA SULTANA

Director (Programs)(Retd), Environment Protection Training and Research Institute (EPTRI), Hyderabad

Dr. Razia Sultana, an Indian Scientist, PhD in Environmental Science from Kakatiya University, India. She worked for Andhra Pradesh Pollution Control Board for 14 years and later joined Environment Protection Training and Research Institute (EPTRI) in Hyderabad, and retired as a Director (Programs) after 21 years. She established Biotechnology Laboratory and Botanical Garden at EPTRI. Her professional visits were to most of the states of India, Japan, Sweden, Canada, and United States. Dr. Sultana published papers, and has been a guide/examiner for Bachelor, Master, MPhil and PhD students. She chaired/presented technical papers in national and international conferences. Dr. Sultana assisted the Andhra Pradesh Government in bringing out special legislation on Water, Land and Trees Act, 2000 and organized training programs, executed environmental projects, and conducted research for national and international agencies. She also worked as Course Director for International Institute for Waste Management (IIWM), Bhopal. She served as a member of Environment Panel for Confederation of Indian Industries. She published quarterly Newsletters under Environment Information System (ENVIS) sponsored by Ministry of Environment and Forests, Govt. of India.

DR. SRI HARSHA KOTA

Dr. Sri Harsha Kota is currently an assistant professor of Civil Engineering at Indian Institute of Technology, Guwahati, India. He received his master's and PhD degrees in Civil Engineering from Texas A&M University, College Station, USA. His research is focused on developing emission inventories from mobile sources and landfills, urban air quality monitoring and modeling, noise pollution, developing indoor air purification systems, regional air quality modeling, and climate change. He is an author of several journals and international conference publications. He is currently involved in many funded research projects. He was an invited speaker for several technical talks, and session chair in an international conference. He was in the advisory committee for an international conference and the chief coordinator of a national workshop on air quality modeling. He is the guest editor for a special issue in the *Journal of Environmental Research.*

PREFACE

The book presents the fundamentals of solid and hazardous waste management in a lucid manner. The technologies adopted for efficient treatment, effective storage, and safe disposal of municipal, plastic, biomedical, industrial, hazardous, and electronic wastes are dealt with in great detail. Added to this, Soil remediation technologies, Waste Minimization and Environmental Impact Assessment are also dealt with.

The last chapter deals with Landfill gases. The factors affecting landfill gases generation were discussed in this chapter. Techniques to estimate methane from landfill both on field and in laboratory are described. The models used in different countries to estimate methane emissions from landfill are also discussed at length.

Waste management is an essential task, which has important consequences for public health and well-being, the quality and sustainability of urban environment, and the efficiency and productivity of the urban economy.

In recent years, systems for transfer, recycling, and disposal of solid waste are being updated by many bilateral and multilateral development agencies. This called for an increasing concern for capacity building at the level of municipal management. With its broad organizational implications and close links to other sectors, Municipal Solid Waste Management constitutes an important entry point for integrated urban management support.

Biomedical wastes, hazardous wastes, industrial solid wastes, plastic wastes, and e-wastes are also posing problems of collection, transport, treatment, and disposal.

This book is meant to meet the requirements of Civil and Chemical engineering students opting for specialization in environmental engineering and sciences as well as students of environmental economics and related sciences.

The technologies described in the book will be of great use to decision-makers in municipalities, hospitals, industries, regulatory bodies, planning and government departments in designing systems for safe disposal of municipal solid wastes, plastic waste, biomedical wastes, hazardous wastes, and e-wastes.

It is hoped that the book enjoys patronage from students, teachers, consultants, engineers, scientists, Pollution Control Boards/Pollution

Control Committees, hospitals, industries, NGOs, and people engaged in the area of waste management.

The authors referred to many works during the course of this work and the references given at the end would be very helpful to the readers desiring more information on the subject. The authors are grateful to all those who helped them directly and indirectly in bringing out this book.

The authors are also thankful to Prof. M.A. Singaracharya, Kakatiya University, Dr. K.S.M. Rao, Ramky, Mr. P. Parthasarathi, E-Parisara, and Mr. Amjad Ansari, G J Multiclave for permitting to use material and photographs in the text.

CHAPTER 1

Introduction

Problems associated with the management of solid and hazardous waste in developing countries are diverse and complex. Rapid development, urbanization, and population growth, problems pertaining to the effective management of solid waste, have escalated to a great extent. There are five distinct facets to the management of solid wastes:

1. identifying and categorizing the source and nature of the waste
2. segregation, storage, and collection of the waste
3. transport of the waste
4. processing (including resource recovery) of the waste
5. ultimate disposal of the waste

Until recently disposal was the only technical and economical option that could be taken in the management of these wastes. The technology for resource recovery and recycling was not considered because of its economical impracticability and its low quality for use as raw material in production. However, the time has come when recycling needs to be considered as a strong alternative against disposal for the reasons that the present waste management techniques are not adequate to prevent environmental pollution. Technology development is key to ensure environmental protection and legal compliance. Environmental policy includes the will to use cleaner technologies or best-available technologies. Skills and competences of the labor force have to be sufficiently updated and balanced with technology used. Objectives and targets have to be designed taking into account environmental performance (planned and actual) and technology changes. Operational control and noncompliance management require a periodic assessment and review of the technological means. The auditing process and team shall correspond to the technology development of the organization. Several environmental technologies are presented, covering the following issues:

1. Municipal solid waste
2. Plastic waste
3. Biomedical waste
4. Hazardous waste
5. Electronic waste

Solid and Hazardous Waste Management.
DOI: http://dx.doi.org/10.1016/B978-0-12-809734-2.00001-8

Waste management is one of the most important environmental problems of the world. Various different technologies exist to apply to management of the waste that human activities generate. The best option to combat the waste accumulation problems is always a reduction in generation of wastes, then the reuse of wastes, and finally the recycling of wastes. Recent advancement is to recover. Sometimes it is necessary to treat and dispose the wastes.

The techniques used to manage wastes are of three types:

- volume reduction technologies (mechanical, physical, and chemical)
- treatment and disposal of waste technologies (biodegradation, solidification, stabilization, disinfection, destruction, dismantling, etc.)
- ultimate disposal of wastes

Quantification, characterization and categorization, and technologies to manage the above wastes are presented in the following chapters.

CHAPTER 2

Municipal Solid Waste

2.1 INTRODUCTION TO MUNICIPAL SOLID WASTE

Municipal solid waste includes commercial and residential wastes generated in a municipal or notified area in either solid or semisolid form excluding industrial hazardous wastes but including treated biomedical wastes.

The amount of municipal solid waste generated in various cities varies from 0.3 to 0.6 kg/capita/day. It has been estimated that there is potential for preparing 8 million tonnes of agricultural manure per year from urban waste in India. At present, less than 25% of the available potential is being exploited. Disposal of urban refuse by composting is a practical solution since it not only takes care of the sanitation problems but also provides a useful agricultural input in the form of soil conditioners as nitrogen, phosphorus, and potassium (NPK) nutrients. Semimechanical plants, where composting is in windrows, is more suitable for Indian conditions. There are several composting plants in different cities, e.g., Hyderabad, Delhi, Mumbai, Ahmedabad, Chandigarh, etc. A great deal of research is being done in the country on the recovery of energy from solid wastes. The costs of garbage and rubbish disposal often exceed 20 percent of the municipal budgets of cities. There is an urgent need to reduce these costs while at the same time extending the levels of services throughout urban areas. This may be accomplished through integrated systems for resource recovery and reuse, in which existing waste disposal and recycling practices are extended and optimized.

There are many impediments to the full adoption of Western technology for a solution to the problems of India:

1. Wastes generated in the developing countries tend to be of low calorific value, high in organic putrescible content and moisture, and are subject to seasonal variations.
2. In the tropical region there are sudden climatic changes, which have to be accounted for in planning solid waste management schemes.

Solid and Hazardous Waste Management.
DOI: http://dx.doi.org/10.1016/B978-0-12-809734-2.00002-X

3. Municipal solid waste disposal costs often exceed 20 percent of municipal budgets. Labor and energy absorb the major portion of the operation costs. More than one percent of the national workforce may be employed in these tasks, and in these services absorb up to one percent of the nation's gross national product (GNP). Thus solid waste management is one of the most expensive services, and systems must be tailored to financial capacity.
4. Solid waste management costs are comprised of four main elements:
   a. capital expenditure and transport facilities
   b. operating costs in the form of oil or power
   c. capital expenditure on buildings
   d. operating expenses on labor

The cost of the first two items is determined by manufacturing costs in industrialized countries and by the prevailing price of oil. Their financial impact is very severe.

In conclusion, the ideal solution is that which results in the maximum reduction in the generation of waste by way of recovery and reuse. The following four approaches to the solution can be thought of:

1. attraction of recycling in its basic form, if the refuse contains valuable reusable materials;
2. reuse as a job creation program;
3. generation of valuable refuse compost or other products; and
4. cost reduction through use of appropriate technologies.

2.2 ORGANIZATION AND MANAGEMENT OF MUNICIPAL SOLID WASTE

Municipal solid waste is defined to include refuse from households, non-hazardous solid waste from industrial, commercial, and institutional establishments (including hospitals), market yard waste, and street sweepings. Semisolid wastes such as sludge and night soil are considered to be the responsibility of liquid waste management systems. While hazardous industrial and medical wastes are, by definition, not components of municipal solid waste, they are normally quite difficult to separate from municipal solid waste, particularly when their sources are small and scattered. Municipal solid waste management (MSWM) systems should therefore include special measures for preventing hazardous materials from entering the waste stream and, to the extent that this cannot be ensured, alleviating the serious consequences that arise when they do.

Finally, debris from construction and demolition constitute "difficult" categories of waste, which also require separate management procedures.

Management is a cyclical process of setting objectives, establishing long-term plans, programming, budgeting, implementation, operation and maintenance, monitoring and evaluation, cost control, revision of objectives and plans, and so forth. Management of urban infrastructure services is a basic responsibility of the municipal government. It is usually advantageous to execute service provision tasks in partnership with private enterprises (privatization) and/or with the users of services (participation), but the final responsibility remains that of the government. MSWM refers to the collection, transfer, treatment, recycling, resource recovery, and disposal of solid waste. The goals of MSWM are:

- to protect environmental health
- to promote the quality of the environment
- to support the efficiency and productivity of the economy
- to generate employment and income

To achieve the above goals, it is necessary to establish sustainable systems of solid waste management that meet the needs of the entire urban population, including the poor. The essential condition of sustainability implies that waste management systems must be absorbed and carried by the society and its local communities. These systems must, in other words, be appropriate to the particular circumstances and problems of the city and locality, employing and developing the capacities of all stakeholders, including the households and communities requiring service, private sector enterprises and workers (both formal and informal), and government agencies at the local, regional, and national levels. Waste management should be approached from the perspective of the entire cycle of material use, which includes production, distribution, and consumption, as well as waste collection and disposal. Whilst immediate priority must be given to effective collection and disposal, waste reduction and recycling should be pursued as equally important, longer-term objectives. The principles of sustainable waste management strategies are thus to:

- minimize waste generation
- maximize waste recycling and reuse
- ensure the safe and environmentally sound disposal of waste

Solid waste management goals cannot be achieved through isolated or sectoral approaches. Sustainable waste management depends on the overall effectiveness and efficiency of management, and the capacity of responsible municipal authorities. Within the overall framework of urban management, the scope of MSWM encompasses detailed planning and management wherein strategic planning, legal and regulatory frameworks, public participation, financial management (cost recovery, budgeting, accounting, etc.), institutional arrangements (including private sector participation), and disposal facility sites are various functions and concerns. The waste generation and characteristics of waste should be viewed with regards to its source of generation and composition, etc. Waste handling is a function concerning collection, transfer, treatment, and disposal, also taking care of special wastes such as medical and small industrial waste, etc.

Practical strategies for improving MSWM will thus comprise specific objectives and measures in these areas. Actors and partners are a wide range of individuals, groups, and organizations that are concerned with MSWM as service users, service providers, intermediaries, and/or regulators. The interests, agendas, and roles of these actors are briefly described below.

Households, Communities, and Other Service Users: Residential households are mainly interested in receiving effective and dependable waste collection service at a reasonably low price. Disposal is not normally a priority demand of service users, so long as the quality of their own living environment is not affected by dumpsites. Only as informed and aware citizens do people become concerned with the broader objective of environmentally sound waste disposal. In low-income residential areas where most services are unsatisfactory, residents normally give priority to water supply, electricity, roads, drains, and sanitary services. Solid waste is commonly dumped onto nearby open sites, along main roads or railroad tracks, or into drains and waterways. Pressure to improve solid waste collection arises as other services become available and awareness mounts regarding the environmental and health impacts of poor waste-collection service. Poorly served residents often form community-based organizations to upgrade local environmental conditions, improve services, and/or petition the government for service improvements. Community-based organizations, which may arise in middle- and upper-income neighborhoods as well as in low-income areas, may become valuable partners of the government in local waste management. When sufficiently organized, community groups have considerable potential for managing and financing local collection services and operating waste recovery and composting activities.

Small- and Large-Scale Industries, Commercial Establishments, Institutions, and Other Service Users: They are similarly interested in reliable and affordable waste collection service. Commercial establishments are particularly concerned with avoiding waste-related pollution, which would inconvenience their customers. Industrial enterprises may have a strong interest in reducing waste generation and can play an active role in managing waste collection, treatment, and disposal in collaboration with government authorities and/or specialized private enterprises.

Nongovernmental Organizations: Nongovernmental organizations (NGOs) operate between the private and governmental realms. Originating outside of the communities in which they work, NGOs are motivated primarily by humanitarian and/or developmental concerns rather than an interest in service improvement for their own members. The self-creation of meaningful employment for members may also be a motivation for NGO formation. NGOs may help increase the capacity of people or community groups to play an active role in local solid waste management by contributing to:

- people's awareness of waste management problems
- organizational capacity and the formation of community-based organizations
- channels of communication between community-based organizations and government authorities
- community-based organizations' voice in municipal planning and implementation processes
- technical know-how of locally active community-based organizations
- access to credit facilities

NGOs may also provide important support to informal-sector waste workers and enterprises, assisting them to organize themselves, to improve their working conditions and facilities, increase their earnings, and extend their access to essential social services such as healthcare and schooling for children.

Local Government: Local government authorities are generally responsible for the provision of solid waste collection and disposal services. They become the legal owner of waste once it is collected or put out for collection. Responsibility for waste management is usually specified in bylaws and regulations and may be derived, more generally, from policy goals regarding environmental health and protection. Besides their legal obligations, local governments are normally motivated by political interests. User satisfaction with provided services, approval of higher government authorities, and financial viability of the operation are important criteria for successful solid waste management from the local government perspective.

The authority to enforce bylaws and regulations, and to mobilize the resources required for solid waste management, is, in principle, conferred upon local governments by higher government authorities. Problems often arise when local governments' authority to raise revenues is not commensurate with their responsibility for service provision. Besides solid waste management, municipal governments are also responsible for the provision of the entire range of infrastructure and social services. Needs and demands for MSWM must therefore be weighed and addressed in the context of the needs and relative priorities in all sectors and services. To fulfill their solid waste management responsibilities, municipal governments normally establish special-purpose technical agencies, and are also authorized to contract private enterprises to provide waste management services. In this case, local authorities remain responsible for regulating and controlling the activities and performance of these enterprises.

Effective solid waste management depends upon the cooperation of the population, and local governments should take measures to enhance public awareness of the importance of MSWM, generate a constituency for environmental protection, and promote active participation of users and community groups in local waste management.

National Government: National governments are responsible for establishing the institutional and legal framework for MSWM and ensuring that local governments have the necessary authority, powers, and capacities for effective solid waste management. The responsibility is delegated without adequate support to capacity building at the local government level. To assist local governments to execute their MSWM duties, national governments need to provide them with guidelines and/or capacity-building measures in the fields of administration, financial management, technical systems, and environmental protection. In addition, national government intervention is often required to solve cross-jurisdictional issues between local government bodies, and to establish appropriate forms of association.

Private Sector Enterprises as Service Providers: The formal private sector includes a wide range of enterprise types, varying from informal microenterprises to large business establishments. As potential service suppliers, private enterprises are primarily interested in earning a return on their investment by selling waste collection, transfer, treatment, recycling, and/or disposal services. Operating in various forms of partnership with the public sector, they may provide capital, management and organizational capacity, labor, and/or technical skills. Due to their profit orientation, private enterprises can, under appropriate conditions, provide MSWM services more effectively and at lower costs than the public sector. However, private sector involvement does not, in itself, guarantee effectiveness and low costs. Problems arise when privatization is poorly conceived and regulated and, in particular, when competition between suppliers is lacking. Private sector waste collectors may be contracted directly by individual households, neighborhood associations, or business establishments. More often, they operate under contractual agreement with municipal authorities. In this case, the authorities commonly retain responsibility for user-fee collection. This arrangement ensures more equitable service access; when private enterprises depend on the direct collection of user charges they have little incentive to provide services in low-income areas where revenue potentials are weak.

Informal Private Sector as Service Providers: The informal private sector comprises unregistered, unregulated activities carried out by individuals, families, groups, or small enterprises. The basic motivation is self-organized revenue generation; informal waste workers are often driven to work as waste collectors or scavengers by poverty and the absence of more attractive employment possibilities. In some cases, informal waste workers belong to religious, caste, or ethnic minorities and social discrimination is a factor that obliges them to work under completely unhygienic conditions as waste collectors or "sweepers." Informal waste workers usually live and work under extremely precarious conditions. Scavenging, in particular, requires very long working hours and is often associated with homelessness. Besides social marginalization, waste workers and their families are subject to economic insecurity, health hazards, lack of access to normal social services such as healthcare and schooling for children, and the absence of any form of social security.

The waste collection, transfer, separation, recycling, and/or disposal activities of informal waste workers constitute economically valuable services. Informally organized groups, in some cases, are hired directly by households and/or neighborhood groups. In general, however, the marginalized and unstable social and economic circumstances of informal waste workers make it quite difficult to integrate their contribution into the MSWM system. As an initial step, informal workers require organizational and technical support to promote their social rehabilitation and alleviate the unacceptable socioeconomic conditions in which they live and work. Through the formation of cooperative societies or microenterprises, it is often possible to considerably increase the job stability and earnings of informal sector workers, and to enhance the effectiveness of their contribution to waste management.

External Support Agencies: Numerous bilateral and multilateral external support agencies are engaged in supporting MSWM in low-income countries. While some external support agencies have acquired considerable expertise in the area of waste management, there is a pressing need to improve cooperation between external support agencies active in the field of MSWM. Due to a lack of coherence in the technical and developmental concepts of successive external support agencies' contributions, many cities in developing countries are encumbered with incompatible and ineffective MSWM facilities and equipment. Coordination of approaches and activities would also enhance the effectiveness of external support agencies' contributions at the national and regional levels. Besides multi- and bilateral development agencies, coordination should encompass external NGOs working in areas related to waste management.

The effectiveness and sustainability of MSWM systems depend upon their adaptation to the prevailing context of the city and/or country in which they operate. The most important aspects in this respect are outlined below at the political, sociocultural, economic, and environmental levels.

Political Context: MSWM is influenced in numerous ways by the political context. The existing relationship between local and central governments (e.g., the effective degree of decentralization), the form and extent of citizen participation in the public processes of policy making, and the role of party politics in local government administration all affect the character of management, governance, and the type of MSWM system that is possible and appropriate.

Sociocultural Context: The functioning of MSWM systems is influenced by the waste-handling patterns and underlying attitudes of the population, and these factors are, themselves, conditioned by the people's social and cultural context. Programs to disseminate knowledge and skills, or to improve behavior patterns and attitudes regarding waste management, must be based on sound understanding of social and cultural characteristics. Fast-growing low-income residential communities may comprise a considerable diversity of social and ethnic groups, and this social diversity strongly influences the capacity of communities to organize local waste management. The effectiveness and sustainability of municipal waste management systems depends on the degree to which the served population identifies with and takes ownership of the systems and facilities. To this end, it is important that the people be involved from the outset in the planning of the local segments of waste management systems. Community involvement is particularly important regarding the siting of facilities such as waste transfer stations and landfill sites.

Economic Context: Waste management tasks, and the technical and organizational nature of appropriate solutions, depend on the economic context of the country and/or city in question and, in fact, on the economic situation in the particular area of a city. The level of economic development is an important determinant of the volume and composition of wastes generated by residential and other users. At the same time, the effective demand for waste management services, and willingness and ability to pay for a particular level of service, are also influenced by the economic context of a particular city or area.

Environmental Context: Firstly, the size and structure of a settlement has an important influence on the urgency of waste management needs. In quite low-density semiurban settlements, some form of local or even onsite solution to the management of organic solid wastes may be more appropriate than centralized collection and disposal. The physical characteristics of a settlement, including such factors as density, width and condition of roads, topography, etc., need to be considered when selecting and/or designing waste collection procedures and equipment such as containers and vehicles. Secondly, at the level of natural systems, the interaction between waste-handling procedures and public health conditions is influenced by climatic conditions and characteristics of local natural and ecological systems. The degree to which uncontrolled waste dumpsites become breeding grounds for insects, rodents, and other disease vectors and a gathering place for dogs, wild animals, and poisonous reptiles depends largely on prevailing climatic and natural conditions. In practical terms, climate determines the frequency with which waste collection points must be serviced in order to limit negative environmental consequences.

Environmental health conditions may also be indirectly affected through the pollution of ground and surface water by leachates from disposal sites. Air pollution is often caused by open burning at dumps, and foul odors and wind-blown litter are common. Methane, an important greenhouse gas, is a byproduct of the anaerobic decomposition of organic wastes in landfill sites. In addition, waste dumps may also be a source of airborne bacterial spores and aerosols. The suitability of a disposal site depends upon many factors, including specific characteristics of the subsoil, ground water conditions, topography, prevailing winds, and the adjacent patterns of settlement and land use.

In MSWM, strategic, political, institutional, financial, economic, and technical aspects play a major role, the details of which are as follows.

Strategic Aspects: Development cooperation in the field of MSWM aims at establishing sustainable waste management systems. Supported solutions must, in other words, be appropriate to the circumstances, problems, and potentials of the particular city and locality, so that they are absorbed and carried by the municipality and its local communities. A sustainable solution will not necessarily represent the highest standards of service and environmental protection, but those that can be afforded. It is important not to raise inappropriate and unachievable expectations in this regard. At the strategic level, appropriateness means more than passive adaptation to the prevailing context, however. Sustainable strategies of MSWM require that specific objectives must be formulated and appropriate measures taken with regard to the political, institutional, social, financial, economic, and technical aspects of waste management. In practice, MSWM support programs often focus on one or two aspects as entry points. Although it is possible to begin with any one of the above aspects, the sustainability of development strategies will depend upon the eventual engagement of the entire range of these aspects.

The implementation of development strategy is a long-term process involving cooperation and coordination between various actors and partners. Each contribution needs to build upon existing activities and programs, avoiding duplication and promoting linkages and synergy effects between ongoing efforts. Development assistance should enable a "learning-by-doing" approach and promote the dissemination of successful solutions.

Political Aspects: The political aspects of MSWM strategies encompass formulation of goals and priorities, determination of roles and jurisdiction, and establishment of legal and regulatory framework.

Certain goals of MSWM, such as the provision of waste collection service to the poor and the environmentally sound disposal of solid waste, have the character of "public goods," meaning that the total private economic demand for services is considerably lower than the full value of those services to society. In these cases, a public process is required to articulate the full public demand for services and mobilize the corresponding resources. To be politically sustainable, this process must be based on clearly formulated goals that enjoy broad popular support. Under conditions of limited resources and extensive waste management needs, tradeoffs between alternative goals and objectives are inevitable. Society may have to choose between a more extensive coverage of collection services as opposed to higher environmental standards of waste disposal, or between improved waste management as opposed to the upgrading of another infrastructure sector. Governments should also assess the potential for waste minimization and determine what priority should be given to minimization efforts in relation to waste collection and disposal activities. This kind of policy issue cannot be resolved at the technical level alone; it calls, rather, for a consultative, political process of goal formulation and prioritization.

Effective waste management and environmental protection programs call for a clear definition of roles, jurisdictions, legal responsibilities, and rights of the concerned governmental bodies and other organizations. The absence of clear jurisdiction may lead to controversies, ineffectiveness, and/or inaction, undermining the political sustainability of MSWM systems. The potential for establishing effective institutional arrangements for MSWM depends largely on the existing systems of urban and rural planning and administration. As a basis for performance-oriented management, a comprehensive "strategic plan" for the sector is required. This plan should provide relevant quantitative and qualitative information on waste generation and specify targets for waste reduction, reuse, recycling, and service coverage. It should describe the organization of waste collection, transfer and disposal in the medium and long term. Such plans would outline the major system components and the project relationships between various bodies and organizations involved in the system. They would provide guidelines regarding the degree of decentralization of specific waste management functions and responsibilities, the forms of private enterprise involvement in waste management processes, and the role of people's participation. Objectives concerning cost-effective and locally sustainable MSWM would be specified, along with the associated financial policies.

The instrumental basis for implementing the strategic plan comprises a legal and regulatory framework which is elaborated in the form of bylaws, ordinances, and regulations concerning solid waste management and includes corresponding inspection and enforcement responsibilities and procedures at national, state, and local levels. These would also include provisions for the management of industrial and hazardous wastes. Regulations should be few in number, transparent, unambiguous, easily understood and equitable. Furthermore, they should be conceived with regard to their contribution to urban and rural physical and economic development.

Regulation and controls are not the only type of instrument available for achieving waste management goals. Other options include economic incentives, the internalization of externalized costs according to the "polluter pays" principle, and noneconomic motivations based on environmental awareness and solidarity of the population. Authorities should consider the full range of available instruments within the policy framework. The main political objectives are to:

- determine society's goals and priorities for waste management and mobilize public support for these goals,
- achieve a clear definition of jurisdictional arrangements for waste management tasks among the concerned government bodies and private sector actors, as well as the roles, rights, and responsibilities of service users, and
- elaborate an appropriate legal and regulatory framework and body of instruments that enable responsible authorities to achieve and sustain the defined goals.

Institutional Aspects: Institutional aspects of MSWM concern the institutional structures and arrangements for solid waste management as well as organizational procedures and the capacity of responsible institutions. The various institutional aspects are distribution of functions, responsibilities, and authority between local, regional, and central government institutions (decentralization) and among local governments in a metropolitan area. Organizational structure of the institutions responsible for municipal solid waste management should keep in mind the coordination between municipal solid waste management and other sectors of management functions such as

1. procedures and methods employed for planning and management,
2. capacities of institutions responsible for solid waste management,
3. the capabilities of their staff,
4. private sector involvement, and
5. participation of communities and user groups.

Effective solid waste management depends upon an appropriate distribution of function responsibilities, authority, and revenues between national, provincial, and local governments, as well as intraurban entities such as wards or communities. Problems arise when certain functions such as investment programming and revenue collection are centralized, while responsibility for operation and maintenance remains at the local government level. In the wake of metropolitan growth, waste management tasks often extend across several local government units. These circumstances call for "horizontal" cooperation between the municipalities concerned, to achieve an effective and equitable division of MSWM responsibilities, costs, and revenues. Local authorities responsible for solid waste management should be granted authority to manage all related affairs and, in particular, to collect and employ user charges and other revenues for the purpose of MSWM. Decentralization of authority should be accompanied by a corresponding distribution of financial and administrative powers and capacities for system planning, implementation, and operation. This normally requires improved procedures for preparing local solid waste management budgets based on actual costs, and allocating the required funds. Effective decentralization makes solid waste management more flexible, efficient, and responsive to local requirements and potentials. At the same time, decision making, financial management, procurement, and implementation functions reduce the load on the central authorities, allowing them to focus on their main responsibilities in the areas of legislation, definition of standards, environmental monitoring, and support to municipalities.

Decentralization and improved MSWM capacity normally requires innovations in the organizational structures, staffing plans, and job descriptions of responsible local government bodies. Assistance should aim at identifying institutional constraints inherent in the system and increasing competence and autonomy at the local level. Procedures and forms of cooperation between local and central government authorities normally need improvement. In this regard, central government bodies may also require development assistance to enable them to accomplish the shifts in their functions and tasks that are associated with decentralization and to better support local governments in the acquisition of new capacities. The organizational status of the technical agency responsible for solid waste as a municipal department or authority needs to be determined. The appropriate institutional arrangements will vary with the size and developmental status of the city. It may be advisable for large- and medium-sized cities either to establish an autonomous regional or metropolitan solid waste authority, or to delegate collection responsibility to the individual local governments, with the metropolitan authority retaining responsibility for transfer and disposal tasks. In the case of small cities, it may be necessary to provide support for planning and standards development as well as technical and financial assistance from national authorities.

The relationships and linkages between MSWM and other municipal service sectors (sewage and drainage, public works, roads, public health, etc.) need to be clarified within the overall framework of rural and urban management. Finally, the development of municipal-level administrative structures themselves calls for institutional development, elaboration of job descriptions, operational procedures, definition of competencies, etc.

The management approaches, methods, and techniques employed in MSWM are often inadequate. In comparison with other sectors, agencies responsible for solid waste management often pay too little attention to integrated management approaches based on adequate information systems, decentralized responsibility, and interdisciplinary interaction and cooperation between functional levels. Based on the defined role of the local government in MSWM, improvement efforts would give primary attention to appropriate strategic planning and financial management methods, including cost-oriented accounting systems, budget planning and control, unit cost calculations, and financial and economic analysis. With regard to operational planning, appropriate management methods and skills include data collection techniques, analysis of waste composition, waste generation projection and scenario techniques, and formulation of equipment specifications, procurement procedures and management information systems for effective monitoring, evaluation, and planning revision.

Large discrepancies often exist between the job requirements and the actual qualification of the staff at the managerial and operational levels. As an initial step towards improvement, awareness-building measures regarding environmental and sanitation issues may be required among responsible staff. On the basis of the organizational development plan, job descriptions, and training needs analysis, a program for manpower development may be elaborated and an appropriate training program implemented. As appropriate, institutional capability for training and human resources development for MSWM should be established at the city, regional, or country level. Creation of a national professional body for solid waste management may help to raise the profile of the profession and promote improved operational and professional standards.

Private enterprises can usually provide solid waste collection, transfer, and disposal services more efficiently and at lower cost than the public sector. However, formal private sector involvement in solid waste management does not in itself guarantee efficiency. The preconditions for successful private sector involvement include:

- competitive bidding
- existence of enterprises with adequate technical and organizational capacity
- effective regulation of the partnership arrangements
- adequate management of the private partners through clear specifications, monitoring, and control

Private sector involvement in MSWM implies a shift in the principal role of government institutions from service provision to regulation. To effectively regulate and control the activities and performance of contracted private enterprises, appropriate systems of monitoring and control need to be established, and corresponding skills and capacities developed at both local and central government levels. In some cases, it is also advisable to provide technical assistance to those enterprises that demonstrate a potential for engagement in MSWM.

Where municipal waste collection services are insufficient, industrial and commercial establishments occasionally hire private enterprises directly to collect and dispose of their solid wastes, and larger companies sometimes undertake disposal themselves. Both waste generators and private waste management enterprises are interested in reducing costs to a minimum, and this often leads to inadequate waste disposal practices. In this case, the public sector's main task is regulation to ensure that hazardous wastes are separated from ordinary wastes and that both types are disposed of in an environmentally safe manner.

Enhancement of the contribution of informal waste collection workers depends, above all, on improved organization among these workers. Support should aim to:

- improve working conditions and facilities,
- achieve more favorable marketing arrangements for services and scavenged materials (see economic aspects), and
- introduce health protection and social security measures (social aspects).

It is essential that the contribution of informal workers to MSWM be officially recognized and that their activities be integrated into the planning of municipal collection and resource-recovery services.

In the interest of effective service delivery and cost efficiency, solid waste management authorities should seek to establish partnership relationships with residential communities and user groups. Where municipal capacities are inadequate and/or low-cost solutions are essential, responsibility for local collection may be decentralized to the communities themselves. Preconditions for effective participation and community-based waste management include adequate problem awareness and organizational capacities. The support of NGOs may be very useful in building the capacity of communities to participate in local solid waste management. The main objectives at the institutional level are to

- evolve responsibility for MSWM to the local government level and ensure a corresponding decentralization of power and authority,
- establish effective institutional arrangements for waste management at the municipal, and in the case of large cities, at the metropolitan level,
- introduce appropriate methods and procedures that enable efficient waste management services that meet the needs of the entire population,
- build the capacities of municipal institutions and their staff so that they are able to provide the demanded waste management services,
- introduce competition and increased efficiency into solid waste management through the involvement of private sector (formal and informal) enterprises, and
- lower costs and improve the effectiveness of waste management through the participation of communities and service users in local waste management.

The waste generated by a population is primarily a function of the people's consumption patterns and, thus, of their socioeconomic characteristics. At the same time, waste generation is conditioned to an important degree by people's attitudes towards waste, their patterns of material use and waste handling, their interest in waste reduction and minimization, the degree to which they separate wastes, and the extent to which they refrain from indiscriminate dumping and littering. People's attitudes influence not only the characteristics of waste generation, but also the effective demand for waste collection services; in other words, their interest in and willingness to pay for collection services. Attitudes may be positively influenced through awareness-building campaigns and educational measures on the negative impacts of inadequate waste collection with regard to public health and environmental conditions, and the value of effective disposal. Such campaigns should also inform people of their responsibilities as waste generators and of their rights as citizens to waste management services. Attitudes towards solid waste may be positively influenced by public information and educational measures; improved waste-handling patterns can hardly be maintained in the absence of practical waste disposal options. Awareness-building measures should therefore be coordinated with improvements in waste collection services, whether public or community-managed. Similarly, people's waste generation and disposal patterns are influenced by their neighbors. A collective logic is involved, because improved waste-handling practices will only yield significant environmental impacts if most households in an area participate in the improvement. Thus, besides general awareness, improved local waste management depends upon the availability of practical options for waste collection and a consensus among neighbors that improvements are both important and possible.

Finally, industrial establishments present special problems regarding waste disposal patterns due to the volume and/or the occasionally hazardous nature of the generated wastes. Regulation and control measures should be employed as far as possible. However, these measures are seldom very effective when large numbers of small industrial establishments are scattered throughout residential and semiresidential areas. Problem of awareness, reliable service options and consensus are crucial to improving waste generation and disposal patterns of industrial enterprises.

Rapidly growing, informally constructed low-income residential areas present a particular challenge to MSWM. Besides the physical constraints of dense, low-income settlement, the inadequacies of other infrastructure services such as roads, drains, and sanitary facilities often increase waste management problems. The access of collection vehicles or push carts may be difficult where roads and footpaths are unpaved. Existing drains are often clogged with waste materials, and solid waste itself may be contaminated with fecal matter. These conditions lead to a proliferation of vermin and disease vectors and increase environmental health risks.

The interrelated nature of service problems and the active role of residents who are often the builders of their own house call for adapted, sectorally integrated development approaches that depend, to a considerable degree, on the cooperation and participation of residents. Households and community-based organizations have important roles to play, not only as consumers or users of waste collection services, but also as providers and/or managers of local-level services. In many low-income residential areas, community-based solid waste management is the only feasible and affordable solution. The introduction of community-based solutions calls for awareness-building measures as well as organizational and technical support. Local NGOs and community leaders may provide essential input towards building community capacity for waste management. Particular attention needs to be paid to the role of women, who normally bear principal responsibility for household waste management. While management is often limited to local collection, it may also encompass waste treatment, (e.g., community composting), recovery, and disposal. It is important that community-based collection systems are carefully linked to the municipal system; local collection activities may break down if waste deposited at municipal transfer points tends to accumulate, rather than being transferred to final disposal sites by municipal services.

Even where waste collection services are provided by municipal authorities, user cooperation is essential regarding such factors as proper storage of household waste, waste separation, placement of household containers, and discipline in the use of public collection points. Households and community participation in the proper operation and maintenance of waste collection and disposal systems may be promoted by broadly conceived awareness-building programs dealing with general public health and environmental issues, as well as information campaigns focused on specific MSWM issues. Formal education courses, school programs, dissemination of teaching and learning materials and direct training and motivational programs for Community based Organisation(s) (CBO) and local leaders are effective means for improving awareness and user participation in MSWM.

Participation is important regarding the development of large centralized facilities such as waste transfer stations and landfill sites. While the adjacent residential population may understand the need for such facilities, they would rather have them located elsewhere; this is the common, not-in-my-backyard or NIMBY attitude. Overcoming the NIMBY attitude requires general public understanding of the requirements of waste management, effective communication, and participation of the concerned community in siting decisions.

Informal sector waste workers are often socially marginalized and fragmented. They live and work without basic economic or social security, under conditions that are extremely hazardous to health and detrimental to family, social, and educational development. Support to informal waste workers should aim to improve their working conditions and facilities, increase their earning capacity, and ameliorate their social security, including access to housing, health, and educational facilities. At the same time, the effectiveness of informal workers' contribution to waste collection, recycling, and reuse may be significantly enhanced.

Public sector waste workers and formal private sector workers are also subjected to unhealthy working conditions and poor social security. Access to social and healthcare services should be ensured. Proper equipment and protective clothing can reduce health risks. By contributing to the "professionalization" of the waste worker's role, proper clothing and equipment may also help to alleviate the social stigmatization that is often associated with waste work. The principal social objectives are to orient municipal waste management towards the real service needs and demands of the population, encourage patterns of waste handling and disposal that contribute to the effectiveness and efficiency of municipal waste services, raise the population's awareness of solid waste problems and priorities and promote an effective economic demand (willingness to pay) for waste collection and disposal service, mobilize and support the contribution of communities and user groups to the-self management of local waste collection and disposal services, and to foster their participation in the planning, implementation, and operation of municipal waste management systems.

Financial Aspects: Financial aspects of MSWM include budgeting and cost accounting systems, resource mobilization for capital investments, cost recovery and operational financing, and cost reduction and control.

Adequate budgeting, cost accounting, financial monitoring, and financial evaluation are essential to the effective management of solid waste systems. In many cities, however, officials responsible for MSWM do not have accurate information concerning the real costs of operations. This is often the result of unfamiliarity and/or lack of capacity to use available financial tools and methods. It is sometimes exacerbated by a lack of incentive or even reluctance in the bureaucratic culture of many local administrations to achieve transparency regarding costs and expenditures. Introduction of improved cost accounting and financial analysis should thus be associated with broader efforts to increase the accountability, efficiency, and commercial orientation of municipal infrastructure management. Where accounting expertise is lacking, it may be brought in from the private sector.

The main options available to local governments for financing capital investment in the solid waste sector are resource mobilization for capital investments (local budget resources, loans from financial intermediaries, and/or special loans or grants from the central government). In some countries, municipal bonds may be a workable source of financing. A further option, private sector financing, has attracted increasing interest in recent years. In many countries, though, the central government is and will continue to be the principal source of funding for major infrastructure investments in solid waste and other sectors. It is important, however, that full responsibility for the functions of planning and investment programming remain with the local government, which must subsequently operate and maintain the acquired facilities and equipment. Procedures that facilitate central financing while evolving investment authority and responsibility to the local government (e.g., infrastructure development funds or banks) should therefore be promoted.

To ensure the appropriateness of investment decisions and avoid "white elephants," adequate financial analysis procedures are needed at the local government level at the strategic planning phase.

To recovery cost and operational financing, there are three main options: user charges, local taxes, and intergovernmental transfers. To promote the responsiveness of the supplying agency to user needs and ensure that collected funds are actually applied to waste management, it is usually preferable to finance operations through user charges rather than general tax revenues. Collection efficiency may be increased by adding solid waste utility charges, such as the water bill, where property tax coverage is universal and the municipal government is responsible for its collection. An itemized line on the tax bill may be appropriate.

User charges should be based on the actual costs of solid waste management, and related, as far as possible, to the volume of collection service actually provided. Among larger waste generators, variable fees may be used to manage the demand for waste services by providing added incentive for waste minimization.

While the economic demand for waste collection services may cover primary collection costs, it seldom covers full transfer, treatment, and disposal costs, especially among low-income groups. To achieve equity of waste service access, some cross-subsidization and/or financing out of general revenues will be required. Large-scale waste generators should pay the full cost of disposal services on the "polluter pays" principle.

In practice, municipal government performance in the collection of waste service fees is often quite poor. People are reluctant to pay for municipal waste collection services that are perceived to be unsatisfactory; at the same time, poor payment performance leads to a further deterioration of service quality, and a vicious circle may arise. Improved fee collection can usually be achieved by attaching waste collection charges to the billing of another service such as water supply or electricity. Such systems may be made progressive, in the sense that large users would pay a higher rate per volume of collected waste than small users. In the case of large single-point producers such as industrial or commercial enterprises, volume- or weight-based charges may be more appropriate; this has the advantage of linking waste revenues to the actual volume of services provided.

In many cities, solid waste service revenues flow into a general municipal account, where they tend to be absorbed by overall expenditures instead of being applied to the intended purpose of waste management. The danger of such misallocation of funds is even greater when locally collected fees and revenues are transferred to the central government before being redistributed to the local level. Besides the simple fact of reducing funding for waste management, the absence of linkage between revenues and the actual levels of service provision tends to undermine the accountability of local waste management institutions and remove their incentive to improve and/or extend services. Resolution of this problem calls for clear political decisions and autonomous accounting procedures, which ensure that the collected revenues are actually applied.

To ensure the long-term economic sustainability of MSWM systems, investments in system development should correspond to the level of resources that the society can make available for waste management. The potential for increasing revenues from solid waste operations is usually quite limited and the most effective way to ensure financial sustainability is through cost reduction, or "doing more with less." There are almost always opportunities to significantly reduce the operational costs of MSWM services.

In principle, the most straightforward way to lower the variable cost component of waste management is to reduce the waste load at the source, i.e., to minimize the generation of waste. In low-income residential areas, the potential for waste reduction is usually quite limited. Public waste collection costs may be reduced through the participation of residential communities in local solid waste management. In most cases, this involves hiring of small-scale enterprises or informal waste collection workers by CBO. Besides lower-cost collection service, informal waste recovery and/or scavenging also contribute to cost savings by reducing the volume of waste that needs to be transferred and disposed.

Important cost reductions may be achieved by introducing competition through public–private partnerships for waste management. Private enterprises are highly motivated to lower costs and may introduce innovations and efficiency-raising measures to this end. The outcome may be useful for defining realistic performance standards that are also applicable to the public segment of the waste management system.

At the most fundamental level, cost reduction implies a better utilization of available manpower and equipment, improved maintenance of equipment, introduction of appropriate technologies, and the elimination of inefficient bureaucratic procedures. Authorities concerned at the local and central government levels should have access to information on the actual cost of MSWM services and relevant performance standards to better judge the potential for cost reduction. The collection and dissemination of cost data, efficiency indicators, and performance standards may serve to focus managers' attention on those areas of operations that require improvement. The principal financial objectives are to establish practical systems of budgeting and cost accounting for MSWM that yield transparency with regard to the real costs of waste management and provide a basis for planning and improving operational efficiency, mobilize required resources for investment in waste management facilities and equipment, achieve cost-oriented revenues for waste management operations that are based, as far as possible, on user charges, and to ensure that the collected revenues are applied to the intended purpose of waste management and reduce the costs and improve the efficiency of waste management operations.

Economic Aspects: Economic aspects relate to the entire national economy and are primarily concerned with the impact of waste management services on the productivity and development of the national economy, economic effectiveness of waste management systems, conservation and efficient use of materials and resources, and job creation and income generation in waste management activities.

Large- and small-scale industrial activities and commercial activities including shops, markets, hotels, and restaurants are important waste generators. Businesses are obliged to dispose of these wastes that would otherwise encumber their establishment and negatively affect workers, clients, and customers. There is therefore a substantial economic demand for waste collection services from economic activities. As frequently observed, waste generation and the demand for collection services generally increase with economic development.

Efficient, reliable and low-cost MSWM service is vital in the development of the national economy. The objective of lowering service costs may conflict with the goal of environmental protection. To determine the appropriate tradeoff, it is important to obtain accurate and, as far as possible, complete information on the sources and composition of industrial and commercial wastes, including hazardous wastes. Authorities should work closely with private sector firms to devise the best technical, organizational, economical, and environmental solution to the problems of normal and hazardous waste disposal. Considerable efforts at awareness building and technical support are usually required to gain the cooperation of industrial and commercial waste generators. A transparent approach is required, as private enterprises will be very reluctant to pay the extra cost of proper waste handling if they believe that their competitors do not pay.

The overall economic effectiveness of waste collection and disposal service depends on the one hand upon the life cycle costs of facilities, equipment, and services, and on the other hand, on the long-term economic impact of waste management systems. Economic impacts may include such factors as the reduction of illness and healthcare costs, enhancement of environmental quality and property values, reduction of disturbances, and increase of business volumes. The economic evaluation of such factors is in principle an important input to strategic plans and investment programs for developing MSWM systems. Besides their use in the appraisal and justification of investment decisions, economic evaluations may be employed to demonstrate the externalized costs of waste pollution and thus to build popular support for improved waste management. In most cases municipal authorities do not have the capacity to conduct economic evaluation or to tackle the methodological issues involved.

At the macroeconomic level, waste management begins with the efficient use of materials and avoidance of hazardous materials at the phases of production and distribution. Policies should be introduced that restrain wasteful use of materials and encourage waste recovery and reuse. The most effective way to promote material conservation and efficiency is in principle to internalize as far as possible the associated future costs of waste collection and disposal or alternatively, the pollution costs that arise from noncollection in the production, distribution, and consumption phases according to the "polluter pay" principle. Legally obliging producers and/or sellers to take back and safely dispose of used products (e.g., refrigerators, batteries, etc.) is an important means to this end, and should be introduced where practicable for appropriate products. Raising service charges in line (or progressively) with the generated waste volume affects consumer behavior (e.g., packaging materials) and disposal patterns (e.g., waste separation) and may thus be applied to manage demand in the interest of waste minimization. These measures are only effective when applied to high-income areas and/or relatively high-volume waste generators.

Besides reducing costs, privatization of waste management service is also relevant to employment and income generation; in this case, the impact is not necessarily positive. Solid waste management departments often employ large numbers of relatively unproductive workers and private enterprises are able to lower costs and increase efficiency precisely because they manage to "do more with less" to accomplish the same job with fewer workers. In a static situation higher labor productivity (and higher pay) evidently implies a lower number of jobs. However, higher labor productivity and efficiency can also lead to an increase in the number of jobs through the expansion of lower-cost services. Economic strategies should seek to increase labor productivity and efficiency and then generate more revenues and jobs by expanding coverage of lower-cost and efficient services. Experience in the formal and informal private sectors demonstrates that it is possible to significantly increase waste workers' earnings through better facilities and equipment and more productive use of workers' time. The main economic objectives are to promote the productivity and development of the national economy through the efficient provision of waste collection and disposal services for which users are willing and able to pay; ensure the environmentally sound collection, recycling and disposal of all generated waste including commercial waste; ensure the overall economic effectiveness of waste management services through the adequate evaluation of economic costs and benefits; promote waste minimization, materials conservation, waste recovery, and reuse and the long-term efficiency of the economy by practical application of the "polluter (and user) pays" principle; and generate jobs and earnings in waste management activities.

Technical Aspects: Technical aspects of MSWM include the following.

Technical Planning and Design of MSWM Systems: The technical systems established for primary collection, storage, transport, treatment, and final disposal are often poorly suited to the operational requirements of the city or town. In many cases, the provision of imported equipment by international donors leads to the use of inappropriate technology and/or a diversity of equipment types, which undermines the efficiency of operation and maintenance functions.

Solid waste management facilities and equipment should be evaluated, and appropriate technical solutions designed and selected, with careful attention to their operating characteristics, performance, maintenance requirements, and expected life cycle costs. Technical evaluation requires data on waste composition and volumes, indications of important area-specific variations of waste generation and their expected changes over time, an understanding of the disposal habits and requirements of different user groups, and assessment of the technical capability of public and/or private sector organizations responsible for operating and maintaining the systems. Concepts for the progressive upgrading of technical systems should be elaborated within the framework of the strategic plan for MSWM.

Waste Collection Systems: Waste collection systems comprise household and neighborhood (primary) waste containers, primary and secondary collections vehicles and equipment, and the organization and equipping of collection workers, including the provision of protective clothing. Selection of collection equipment should be based on area-specific data on waste composition and volumes, local waste-handling patterns and local costs for equipment procurement and operation and maintenance (labor, fuel, lubricants, tires, etc.).

Regarding the design of local waste collection systems, the most effective results may be obtained through the participation of the concerned communities. Where appropriate, the objectives of material recovery and source separation should be considered. The introduction of source separation must be done in a pragmatic and incremental manner, beginning with pilot activities to assess and encourage the interest and willingness of users to participate.

To extend service coverage, especially in low-income areas, the use of low-cost community-managed primary collection systems should be considered. In the interest of lower costs and efficient operation and maintenance, appropriate, standardized, and locally available equipment should be selected. Design and procurement should be made with close attention to the requirements of preventive maintenance, repair and spare parts availability. The privatization of maintenance and repair may be considered as a means of lowering maintenance costs and optimizing equipment utilization.

Transfer Systems: Transfer systems include temporary waste storage and transfer points, vehicles and equipment for waste transfer, and the procedures for operating and maintaining these facilities and equipment.

Design and expansion of transfer facilities and equipment must match the characteristics of local collection systems and the available capacity of environmentally safe disposal facilities. The size, number, and distribution of transfer stations must be carefully designed to facilitate local collection while achieving efficient transfer operations and minimum transport distances and costs. Detailed cost analysis is required to determine the optimal solution. The technical characteristics and design of transfer points and vehicles must consider the characteristics of local collection systems (hand cart, dumping requirements, etc.). Careful attention must be given to the objectives of reducing local pollution and limiting as far as possible the access of rats and insects. Transfer points are often a choice location for scavengers' activity and arrangements should be explored for accommodating scavenging without accentuating local pollution problems. The selection of vehicles must be based on careful cost analysis that considers transfer ease, haul volume, operation costs, and maintenance requirements.

Waste Recovery and Disposal: In low-income countries, recovery of recyclable materials, mainly paper, glass, metals, and plastics, are normally undertaken by the informal private sector. This economically useful activity should be facilitated by the appropriate design of equipment and facilities for each stage of the collection and disposal process. The effectiveness of informal waste recovery may be further enhanced through active support aimed at improving the organizational capacity of informal workers, improving equipment and facilities for the collection and sorting of materials, and coordinating municipal waste collection and disposal operations with informal recovery. Formal public sector workers often engage in some form of scavenging activity on the site, and it may be necessary to specify the rights and recovery conditions of both formal and informal workers.

The public sector may itself become involved in waste recovery or lease waste recovery rights to formal private sector enterprises. Composting is a most promising area for the recovery of organic materials. Besides reducing the volume of waste that needs to be transferred and disposed, composting generates a valuable soil conditioner for agricultural and horticultural use. Decisions to introduce composting must be market oriented and based on careful economic and financial analysis. Large-scale sector composting operations are seldom financially viable, and the alternative of small-scale decentralized composting plants may be worth considering. In either case, the potential for financially viable composting may be significantly improved through the introduction of waste separation at source. Governments may need to undertake accompanying measures such as the promotion of appropriate household waste storage facilities and information campaigns to encourage waste separation. Alternatively, community-based composting may be promoted. The location of composting operations adjacent to the markets for soil conditioners (e.g., near farms or nurseries) may also bring advantages. Key factors for success include careful attention to product quality, adequate control, and the use of simple technologies.

Other recovery options focus on the energy value of waste materials like incineration and landfill gas utilization. Due to the composition of wastes in many developing countries (high organic and moisture content), and the high investment and operating costs of the sophisticated technology, incineration is rarely a viable option. On the other hand, landfill gas recovery and utilization may be a more promising approach to energy recovery.

Even when waste minimization and recycling are actively practiced there is always a large quantity of waste remaining for disposal in an environmentally sound manner. The authorities should ensure that appropriate sites for new solid waste disposal are made available, and that these sites will become accessible for the timely execution of MSWM improvements. While the technology is fairly simple, landfills involve complex organic processes. To ensure their efficient operation and to limit disturbances and environmental pollution, landfills need to be carefully sited, correctly designed, and well operated. Particular attention must be given to ground water, soil, and air through the control of leachate and gases. Environmental impact assessment, appropriate design criteria and guidelines on recurrent landfill development and operation should be made available to local authorities. Landfill siting is often politically difficult and requires active public information and participation in order to reach a negotiated solution.

The main benefits of properly designed and correctly operated sanitary landfills derive from the discontinuation of current, unacceptable dumping practices and the environmentally sound closure and recovery of existing dumpsites. It is seldom possible to move from open dumping to fully contained sanitary landfill operations in one step. More often a transformation process must be foreseen in which dumping practices are progressively improved and existing sites gradually upgraded. Municipal authorities should be encouraged to start the transformation process rather than wait until it is possible to construct a completely new and appropriately designed landfill facility.

Hazardous and Special Waste Management: Concerted efforts are required to institute and improve environmental monitoring and controls to keep hazardous wastes out of the municipal system, especially landfills, sewers, and drains. Most importantly, potential sources of hazardous materials in industrial wastes, whether they are served by public or private waste collectors, must be identified, registered, and targeted for appropriate management. Although the laws controlling industrial and hazardous wastes are normally enacted at the national and state level, the municipality has the key role in monitoring the generation of industrial and hazardous waste in their areas, identifying suitable sites for environmentally safe disposal, and monitoring the collection and disposal operations. Industrial discharge programs and guidelines on incoming wastes are available to keep hazardous industrial wastes out of sanitary landfills. Special attention must also be given to the management of infectious waste originating from hospitals and other healthcare institutions. The main technical objectives are:

- to achieve optimal life cycle cost effectiveness of solid waste management equipment and facilities, with due consideration of operation and maintenance requirements, operation costs, and dependability;
- introduce coherent technical systems that are adapted to the requirements;
- operations of all concerned actors including service users, informal sector, private enterprises, and public sector waste operations; and
- install and operate technical systems for waste collection, transfer, recovery, treatment, and disposal that reduce local pollution, limit the proliferation of vermin, and protect the environment.

2.3 QUANTITY OF MUNICIPAL SOLID WASTE

Waste generation rates are low in smaller towns whereas they are high in cities over 20 lakh population. The range is between 0.2 and 0.5 kg/capita/day. The population range and average waste generation per capita per day is shown in Table 2.1.

The urban areas' contribution of municipal solid waste generated daily in India is over 70%. The quantity as well as quality of municipal solid waste generated in the metropolitan cities are generally governed by parameters such as population, standard of living, socioeconomic conditions, commercial and industrial activities, food habits, cultural traditions,

Table 2.1 Municipal solid waste generation rate

Population range (in lakh)	Average waste generation per capita per day (kg)
1–5	0.21
5–10	0.25
15–20	0.27
20–50	0.35
50 and above	0.50

Source: NEERI (SWM, Feb, 1996).

climatic conditions, etc. Growing population, commercialization, and industrialization contribute to generation of more waste day by day.

Average municipal solid waste generation in cities in developed countries varies from 1.5 to 3.0 kg/capita/day. In most of the developing countries municipal solid waste is not segregated at source and is found in mixed conditions. There is a small percentage of recyclable material and a larger amount of compostable and inert materials like ash and road dust. There exists an informal sector of rag pickers, who collect recyclable waste from the streets, bins, and disposal sites. They take away paper, plastic, metal, glass, rubber, etc., for their livelihood, but a small quantity of recyclable material is still left behind. Apart from the environmental hazards, improper solid waste management also incurs extra expenditure to the local municipalities, if not tackled properly and optimally. The esthetics of the municipalities also emphasize the efficient management of municipal solid waste. Solid waste can better be treated as a resource and potential raw material for starting new ventures of recycling.

The quantity of waste from various cities (Table 2.2) was accurately measured by the National Environmental Engineering Research Institute (NEERI). The daily quantity was determined on the basis of quantity transported per trip and the number of trips made per day. Forecasting waste quantities in the future is as difficult as it is to predict the changes of waste composition. The factors promoting change in waste composition are equally relevant to changes in waste generation. An additional point, worthy of note, is the change of density of the waste as the waste moves through the management system, from the source of generation to the point of ultimate disposal. Storage methods, salvaging activities, exposure to the weather, handling methods and decomposition, all have their effects on changes in waste density. As a general rule, the lower the level of economic development, the greater the change between generation

Table 2.2 Quantity of municipal solid waste in urban centers

Population range (in millions)	Number of urban centers (sampled)	Total population (in millions)	Average per capita value (kg/capita/day)	Quantity (tonnes/day)
<0.1	328	68.300	0.21	14,343.00
0.1–0.5	255	56.914	0.21	11,952.00
0.5–1.0	31	21.729	0.25	5432.00
1.0–2.0	14	17.184	0.27	4640.00
2.0–5.0	6	20.597	0.35	7209.00
>5.0	3	26.306	0.50*	13,153.00

Source: Background material for Manual on SWM, NEERI, 1996.

and disposal. Increases in density of 100% are common in developing countries, which mean that the volume of wastes decreases by half.

2.4 CHARACTERISTICS OF MUNICIPAL SOLID WASTE

The characteristics of municipal solid waste depend on the source and type of waste generation. The sources of solid waste generation are households; commercial enterprises such as restaurants, hotels, stores, and markets; institutions; slaughterhouses; hospitals; nursing homes; clinics; construction and demolition sites; remodeling and repairing sites; factories; power plants and treatment plants; etc. The various types of waste generated are food waste (waste from the preparation, cooking, and serving of food, market refuse, waste from the handling, storage, and sale of vegetables), rubbish (paper, cardboard, cartons, wood boxes, plastics, glass, rags, clothes, bedding, leather, rubber, grass, leaves, yard trimmings, metals, tins, metal foils, dirt, stones, bricks, ceramics, crockery, bottles, and other mineral refuse), street waste (street sweepings, dirt, leaves, and dead animals), bulky waste (large auto parts, tires, stoves, refrigerators, other large appliances, furniture, and large crates), horticulture waste (tree trimmings, branches, leaves, roadside trees, waste from parks and gardens), biomedical waste (human anatomical waste, animal waste, microbiology and biotechnology waste, waste sharps, medicines and cytotoxic drugs, solid waste, solid waste incineration ash), construction and demolition waste and industrial waste (solid waste resulting from industry process and manufacturing operations, electronic waste, effluent treatment plant and sewage treatment plant sludge etc., ashes and residues, clinkers, hazardous wastes, explosives, radioactive and toxic waste).

Table 2.3 Patterns of composition, physical characteristics and quantities in low-, middle-, and high-income countries

	Low income[a]	Middle income[b]	High income[c]
Composition: (% by weight)			
Metal	0.2–2.5	1–5	3–13
Glass, ceramics	0.5–3.5	1–10	4–10
Food and garden waste	40–65	20–60	20–50
Paper	1–10	15–40	15–40
Textiles	1–5	2–10	2–10
Plastics/rubber	1–5	2–6	2–10
Misc. combustible	1–8	—	—
Misc. incombustible	—	—	—
Inert	—	—	—
Density	20–50	1–30	1–20
(kg/m^3)	250–500	170–330	100–170
Moisture content (% by wt)	40–80	40–60	20–30
Waste generation (kg/capita/day)	0. 4–0.6	0.5–0.9	0.7–1.8

[a]Countries having a per capita income less than US$360.
[b]Countries having a per capita income US$360.
[c]Countries having a per capita income greater than US$3500 (1978 prices).

2.4.1 Physical Characteristics of Municipal Solid Waste

The composition and characteristics (Table 2.3) of municipal solid wastes vary throughout the world. Even in the same country it changes from place to place as it depends on number of factors such as social customs, standard of living, geographical location, climate, etc. Municipal solid waste is heterogeneous in nature and consists of a number of different materials derived from various types of activities. Waste composition varies with socioeconomic status within a particular community, since income determines lifestyle-consumption patterns and cultural behavior. Even then it is worthwhile to make some general observation to obtain some useful conclusions. The major constituents are paper and putrescible organic matter, metal, glass, ceramics, plastics, and textiles; dirt and wood are generally present although not always so, the relative proportions depending on local factors, and the average proportion of constituents reaching a disposal site or sites for a particular urban area changes in the long term, although there may be significant seasonal variations within a year.

Table 2.4 Physical characteristics of municipal solid wastes

Population range (in millions)	Number of cities surveyed	Paper	Rubber, leather & synthetics	Glass	Metals	Total compostable matter	Inert
0.1–0.5	12	2.91	0.78	0.56	0.33	44.57	43.59
0.5–1.0	15	2.95	0.73	0.35	0.32	40.04	48.38
1.0–2.0	9	4.71	0.71	0.46	0.49	38.95	44.73
2.0–5.0	3	3.18	0.48	0.48	0.59	56.67	49.07
>5	4	6.43	0.28	0. 94	0. 80	30.84	53.90

All values are in percent, and are calculated on net weight basis.
Source: Background material for Manual on SWM, NEERI, 1996.

Conclusions may be drawn from this comparative data as the proportion of paper waste increases with increasing national income; the proportion of putrescible organic matter (food waste) is greater in countries of low income than those of high income; variation in waste composition is more dependent on national income than geographical location, although the latter is also significant; and waste density is a function of national income, being two to three times higher in the low-income countries than in high-income countries. Moisture content is also higher in low-income countries and the composition of waste in a given urban center varies significantly with socioeconomic status (household income) (Table 2.4).

A knowledge of the density of a waste, i.e., its mass per unit volume (kg/m^3) is essential for the design of all elements of the solid waste management system viz. community storage, transportation, and disposal. For example, in high-income countries, considerable benefit is derived through the use of compaction vehicles on collection routes, because the waste is typically of low density. A reduction of volume of 75% is frequently achieved with normal compaction equipment, so that an initial density of 100 kg/m^3 will readily be increased to 400 kg/m^3. In other words, the vehicle would haul four times the weight of waste in the compacted state than when the waste is uncompacted. The situation in low-income countries is quite different; a high initial density of waste precludes the achievement of high compaction ratio. Consequently, compaction vehicles offer little or no advantage and are not cost effective.

Significant changes in density occur spontaneously as the waste moves from source to disposal, as a result of scavenging, handling, wetting and drying by the weather, and vibration in the collection vehicles. Density is as critical in the design of a sanitary landfill as it is for the storage,

Table 2.5 Density of municipal solid wastes in some cities at pickup point

Sl. no	City	Density (kg/m³)
1.	Bangalore	390
2.	Baroda	457
3.	Delhi	422
4.	Hyderabad	369
5.	Jaipur	537
6.	Jabalpur	395
7.	Raipur	405

N.B.: The above figures may be taken as indicative and actual field measurements must be made while designing solid waste management schemes for towns and cities.
Source: Solid Waste Management in Developing Countries INSIDOC, 1983

collection and transportation of waste. Efficient operation of a landfill requires compaction of the waste to optimum density after it is placed (Table 2.5).

For bulk density measurement, materials and apparatus needed are a wooden box of 1 m^3 capacity, a wooden box of 0.028 m^3 capacity, and a spring balance weighing up to 50 kg.

Procedure: The solid waste should be taken in the smaller 0.028-m^3 box to give a composite sample from different parts of the heap of waste, then weighed with the help of a spring balance. After weighing, this smaller box (0.028 m^3) is emptied into the bigger 1-m^3 box and the weight of the waste poured into the bigger box is noted. This is repeated till the larger box is filled to the top. The waste should not be compacted by pressure. Fill the 1-m^3 box three times and take the average. Thus, the weight per cubic meter is obtained.

Moisture content of solid wastes is usually expressed as the weight of moisture per unit weight of wet material.

$$Moisturecontent\,(\%) = \frac{\text{Wet weight} - \text{dry weight}}{\text{Wet weight}} \times 100$$

A typical range of moisture contents is 20–45% representing the extremes of wastes in an arid climate and in the wet season of a region having large precipitation. Values greater than 45% are however not uncommon. Moisture increases the weight of solid waste and therefore the cost of collection and transport. Consequently, waste should be insulated from rainfall or other extraneous water.

Moisture content is a critical determinant in the economic feasibility of waste treatment and processing methods by incineration since energy (e.g., heat) must be supplied for evaporation of water and in raising the temperature of the water vapor.

Climatic conditions apart, moisture content is generally higher in low-income countries because of the higher proportion of food and yard waste.

The size distribution of waste constituents in the waste stream is important because of its significance in the design of mechanical separators and shredders and the waste treatment process. This varies widely and while designing a system, proper analysis of the waste characteristics should be carried out.

The calorific value is the amount of heat generated from combustion of a unit weight of a substance, expressed as kcal/kg. The calorific valuc is determined experimentally using a bomb calorimeter in which the heat generated at a constant temperature of 25°C from the combustion of a dry sample is measured. Since the test temperature is below the boiling point of water, the combustion water remains in the liquid state. However, during combustion the temperature of the combustion gases remains above 100°C so that the water resulting from combustion is in the vapor state.

While evaluating incineration as a means of disposal or energy recovery, the following points should be kept in view:

- Organic material yields energy only when dry.
- The moisture contained as free water in the waste reduces the dry organic material per kilogram of waste and requires a significant amount of energy for evaporation.
- The ash content of the waste reduces the proportion of dry organic material per kilogram of waste. It also retains some heat when removed from the furnace.

2.4.2 Chemical Characteristics of Municipal Solid Waste

A knowledge of chemical characteristics of waste is essential in determining the efficacy of any treatment process. Chemical characteristics include (1) chemical, (2) biochemical, and (3) toxic.

Chemical: Chemical characteristics include pH, nitrogen, phosphorus and potassium (N-P-K), total carbon, C/N ratio, calorific value.

Biochemical: Biochemical characteristics include carbohydrates, proteins, natural fiber, and biodegradable factor. The waste may include lipids as well.

Toxic: Toxicity characteristics include heavy metals, pesticides, insecticides toxicity test for leachates (TCLP), etc. (Table 2.6).

Table 2.6 Chemical characteristics of municipal solid wastes

Population range (in millions)	Number of cities surveyed	Moisture (%)	Organic matter (%)	Nitrogen as total N (%)	Phosphorus as P_2O5 (%)	Potassium as K_2O (%)	C/N ratio	Calorific value (in kcal/kg)
0.1–0.5	12	25.81	37.09	0.71	0.63	0.83	30.94	1009.89
0.5–1.0	15	19.52	25.14	0.66	0.56	0.69	21.13	900.61
1.0–2.0	9	26.98	26.89	0.64	0.82	0.72	23.68	980.05
2.0–5.0	3	21.03	25.60	0.56	0.69	0.78	22.45	907.18
>5.0	4	38.72	639.07	0.56	0.52	0.52	30.11	800.70

A knowledge of the classes of chemical compounds and their characteristics is essential in proper understanding of the behavior of waste as it moves through the waste management system. The products of decomposition and heating values are two examples of the importance of chemical characteristics. Analysis identifies the compounds and the percent dry weights of each class. The rate and products of decomposition are assessed through chemical analysis. Calorific value indicates the heating value of solid waste. Chemical characteristics are very useful in assessment of potential of methane gas generation. The various chemical components normally found in municipal solid waste are described below. The product of decomposition and heating values are two examples of the importance of chemical characteristics. Analysis identifies the compounds and the percent dry weight of each class.

The lipids class of compounds includes fats, oils and grease. The principal sources of lipids are garbage, cooking oils, and fats. Lipids have high calorific values, about 38,000 kcal/kg, which makes waste with a high lipid content suitable for energy-recovery processes. Since lipids in the solid state become liquid at temperatures slightly above ambient, they add to the liquid content during waste decomposition. They are biodegradable but because they have a low solubility in waste, the rate of biodegradation is relatively slow.

Carbohydrates are found primarily in food and yard waste. They include sugars and polymers of sugars such as starch and cellulose and have the general formula $(CH_2O)_X$. Carbohydrates are readily biodegraded to products such as carbon dioxide, water, and methane. Decomposing carbohydrates are particularly attractive for flies and rats and for this reason should not be left exposed for periods longer than is necessary.

Proteins are compounds containing carbon, hydrogen, oxygen and nitrogen and consist of an organic acid with a substituted amine group (NH_2). They are found mainly in food and garden wastes and comprise 5–10% of the dry solids in solid waste. Proteins decompose to form amino acids but partial decomposition can result in the production of amines, which have intensely unpleasant odors.

Natural fibers include the natural compounds, cellulose and lignin, both of which are resistant to biodegradation. They are found in paper and paper products and in food and yard waste. Cellulose is a larger polymer of glucose while lignin is composed of a group of monomers of which benzene is the primary member. Paper, cotton, and wood products are 100%, 95%, and 40% cellulose respectively. Since they are highly combustible, solid waste having a high proportion of paper and wood products is suitable for incineration. The calorific values of over-dried paper products are in the range 12,000–18,000 kcal/kg and of wood about 20,000 kcal/kg, which compares with 44,200 kcal/kg for fuel oil.

In recent years, synthetic organic materials (plastics) have become a significant component of solid waste accounting for 5–7%. Plastic being nonbiodegradable, its decomposition does not take place at the disposal site. Besides, plastic causes choking of drains and environmental pollution when burnt under uncontrolled conditions. Recycling of plastics is receiving more attention, which will reduce the proportion of this waste component at disposal sites.

Materials in the noncombustibles class are glass, ceramic, metals, dust, dirt, ashes, and construction/demolition debris. Noncombustibles account for 30–50% of the dry solids.

2.5 EVOLUTION OF MUNICIPAL SOLID WASTE MANAGEMENT

Solid waste management may be defined as that discipline associated with the control of generation, storage, collection, transfer and transport, processing and disposal of solid waste in a manner that is in accordance with the best principle of public health, economics, engineering, conservation, esthetics, and other environmental considerations. The evolution of solid waste management could be traced back to as early as 1880 when it was first practiced in the United Kingdom. Later on, it spread to other parts of the world. The early disposal practices/most common methods of disposal of solid waste were dumping on land, dumping in water, plowing into soil, feeding to hogs, and incineration. Not all these methods were applicable to all types of wastes; for example, plowing into the soil was used for food wastes and street sweepings, etc.

Dumping on land was a common practice, involving hauling the solid waste to the edge of the town and dumping it there; open dumps became a common method of disposal and burning on these dumps was a common practice. Open dumps also attracted files and rats that spread diseases, hence open dumps lead the way for disease-free sanitary landfill practice.

The method of dumping in water was practiced by some coastal cities until the 1930s, when the pollution consequences of such practices were finally recognized.

The method of plowing into the soil was used for the disposal of food wastes and street sweepings. But due to large land requirement and separation of food wastes, this method was not used extensively.

Feeding to hogs was practiced until it was recognized that hogs were infected by a disease known as trichinosis, which in turn infects human beings consuming hogs. Hence, since 1950 this method has been avoided.

Although incineration was considered to be a final disposal method, it is now considered to be either a volume reduction or an energy conservation process.

2.6 MUNICIPAL SOLID WASTE MANAGEMENT

The problems associated with management of solid waste in today's society are complex because of the quantity and diverse nature of the wastes, the development of sprawling urban areas, the funding limitations for public services in many large cities, the impacts of technology, and the emerging limitations in both energy and raw materials. Hence, for efficient solid waste management, activities associated with management of solid wastes from the point of generation to final disposal have been grouped into six functional elements:

1. waste generation
2. onsite storage
3. door-to-door collection
4. transfer and transport
5. processing and recovery
6. disposal

One of the goals of solid waste management is the optimization of these systems to provide the most efficient and economic solution, commensurate with all constraints imposed by the system and those affected by it or controlling its use. The useful framework of solid waste management could be established by considering the functional elements separately as described below.

Waste generation includes those activities in which the materials are identified as of no value and are either thrown away or collected for disposal. Though at present waste generation is not a functional element, in future it may be one due to the utility of recovered waste such as newspaper, cardboard, aluminum cans and bottles, etc.

The onsite storage place is where solid waste that is heterogeneous in nature will be stored in areas with limited storage space, where people live. These wastes cannot be tolerated at individual premises because of their biodegradability and must be removed within a reasonable time. The cost of providing storage for solid waste at source is borne by householder or apartment owner or by the management of commercial and industrial properties.

The door-to-door collection includes gathering of solid wastes from households/onsite storage and hauling them to the location where the collection vehicle is emptied. The hauling of wastes in small cities is not a problem as disposal sites are nearby but is a problem in large cities as the disposal sites are far off.

Transfer and transport involves two parts, namely the transfer of wastes from the collection vehicle to the larger transport equipment and subsequent transport of the wastes to the processing and recovery site or to the disposal site. Motor vehicle transport is commonly used for handling of solid wastes.

The functional element of processing and recovery includes all the techniques, equipment and facilities used both to improve the efficiency of the other functional elements and to recover usable materials, conversion products, or energy from solid wastes. The recovery of materials includes size reduction, density separation by air classifiers, magnetic devices to pull out iron, current separators for aluminum, and screens for glass. The cost of separation versus the value of recovered materials determines the selection of any recovery process. Processing of waste may produce either fuel or compost.

Disposal is the ultimate fate of all types of wastes. Also incinerator residue, or other substances from the various solid waste processing plants that are of no further use, requires disposal. A modern landfill (sanitary) is used for the final disposal of solid wastes in such a way so as to avoid nuisance or hazards to public health, breeding of rats and insects, and contamination of ground water, etc. Engineering principles must be followed to confine the wastes to the smallest possible area to reduce them to the lowest practical volume by compaction at the site, and to cover them after each day's operation to reduce exposure to the environment. These landfill areas should not be used for buildings as there may be uneven settlement due to decomposition of organic matter but can be used for golf courses, parks, athletic fields, etc.

2.7 TECHNOLOGIES FOR MUNICIPAL SOLID WASTE MANAGEMENT

Various technologies for solid waste management may be rationalized under the following headings:

1. processes for final disposal either of all the waste or of any residue remaining after treatment
2. treatment to achieve volume reduction prior to final disposal
3. separation of the organic from the inorganic fraction of the waste
4. recovery of materials from the inorganic fraction
5. recovery of materials from organic fraction
6. reclamation of organic fraction to produce either a fuel or a chemical product

The major technologies currently in use for municipal solid waste disposal are landfill, composting, and incineration.

2.7.1 Landfill

Sanitary landfill is a method of disposing of refuse on land without creating nuisance or hazards to public health or safety, by utilizing the principles of engineering to confine the refuse to the smallest practical area, to reduce it to the smallest practical volume and to cover it with a layer of the earth at the conclusion of each day's operation or at such more frequent intervals as may be necessary. Thus, the method essentially consists of laying the material systematically followed by its compaction to the smallest practical volume with the least exposed area and then covering it with soil. As the exposed surface area will be smallest, the amount of soil cover needed will be small. Covering of the waste with soil or other inorganic material makes it inaccessible to flies and rodents and the heat released during decomposition is conserved, increasing the chances of the destruction of fly larva and pathogenic organisms. To suit to different site conditions, the basic process of landfilling is divided into three different types, namely the trench method, area method, and ramp method.

The trench method is best suited for that land where excavation can be carried out easily and where the ground-water table is sufficiently low. A trench 2 m deep and 2–5 m wide is cut. The length of the trench depends on site conditions and number of trucks likely to arrive simultaneously, and as such that it takes a day's refuse quantity. The excavated soil is later used to give soil cover.

The area method is best used in the area where natural depressions exist as in quarries and valleys. The waste is put in the natural depressions and compacted. A layer of earth is given on top and compacted. The process is repeated till the depression is filled up. The earth has to be excavated from borrow pits at the site itself or imported from elsewhere.

The ramp method is a modified form of the above two methods. A ramp of about 15 m wide, 30 m long, and of suitable height is created. A shallow cut is taken at the foot of the ramp and a valley-like trench is cut so that the tractor can operate transversely across its width. Trucks come to the top of the ramp and discharge their contents inside the trench simultaneously. Then the refuse is compacted by tractors. The thickness to which a layer can be laid and compacted before giving soil cover depends on the ease of operation of mechanical equipment. The newly laid layers are not completely consolidated; the risk increases with the thickness of layer. Hence, the thickness of layer is restricted to 2 m.

The total compaction and settlement consists of primary consolidation, secondary compression, creep, and decomposition. In the first stage a large portion of settlement occurs in short duration and is also known as short-term shear deformation; the second stage proceeds slowly. As the organic matter after decomposition is converted to stable end-products, the resultant increase in density is reflected by further settlement. Out of three stages, the second and third stages are slow and cannot be mechanically hastened. Primary consolidation depends on weight, composition, and arrangement of particles, depth of fill, and moisture penetration. Of these factors, only unit weight of fill material can be changed. This increase is achieved by using heavy equipment, which due to large static compactive force and dynamic forces, results in better arrangement of particles.

The mechanical equipment at landfill sites are used for the following purposes:

- leveling of waste
- compaction
- excavated and conveyance of soil for cover

The normal practice is to use two sets of equipment, one of which performs two functions. A truck-type bulldozer of the low ground-pressure type can level the material as well as provide compaction. Due to its slow speed, it can operate economically over short distances, up to 100 meters. A Caterpillar D4-type bulldozer can handle about 200 tonnes of refuse in 8 hours of operation. The useful life of such equipment is about 10,000 working hours. It is desirable to use landfill blades on such bulldozers. The truck-type unit distributes its load over a larger area and hence is more stable than a wheel-type unit.

The front-end loaders, which have a hydraulically operated bucket of 0.5–3 cm^3 capacity, can be used for leveling of deposited solid waste and for transferring soil from borrow pit to the working face. The scraper can be self-propelled or towed by a tractor and has a cutting edge that removes a thin soil layer, which is stored in its body.

As a result of natural and artificial rearrangement of particles, the densities of landfill sites increase. The final in situ densities that can be attained depend on the characteristics of solid wastes.

The manual method of landfilling is usually practiced in developing countries such as India, as the solid wastes in such countries do not contain bulky wastes such as furniture, etc. Also, the density of waste in landfill sites of India is found to be higher (300–600 kg/m^3) than that observed in landfill sites of developed countries (125–200 kg/m^3). The following steps are to be followed in the manual method:

- Selection of site should be made using the same criteria as for the mechanized method.
- Provide an all-weather access road from an existing main road to the point at which filling is to commence. Such a road can be prepared from the construction and demolition waste, ash, or clinker, and a small stock of this material should be kept for day-to-day repairs.
- To help guide vehicles to the spot, provide flags or pegs on the location.
- To indicate height to which filling has to be done, "sight rails" should be provided.

- The filling should start from a point nearest to the road, with vehicles approaching the point after reversing. Tipping vehicles can unload faster and assure a quicker turnout. The dumped material can be spread and leveled manually using rakes having a number of teeth. By using the ramp method, the filling will move progressively inside the site.
- To indicate the point where the vehicle should stop for unloading, a strong heavy wooden bumper bar can be provided.
- To keep the rear wheels of vehicles from sinking in the newly deposited mass, cover the area near working face with steel or wooden sleepers.
- Cover the waste at the end of daily operation.

The manual method needs about 50–60 workers per million population.

Land requirement, land use restriction, approach to the site, haul distance, availability of cover material, hydrological investigations, presence of water bodies, etc., are various points to be considered while selecting a site for the landfill. Environment-impact assessment studies should be carried out before finalization of landfill site. The volume of fill required depends upon density, degree of compaction, depth of fill and life for which the site is to be used. The volume required will change in different cases. At a waste generation rate of 0.33 kg/capita/day and final density of 1000 kg/m^3, about 15,000 m^3 will be needed per million population for one year's operation. To know the land-use restrictions the town-planning authorities should be consulted before selecting a particular site so that it is compatible with their plans. With regard to approach, the site should be easily accessible for vehicles throughout the year. The site should not be too close to residential and commercial localities. Provided all other conditions are satisfied, the site should be as near the area to be served as possible, so that transportation cost is reduced. If the soil cover is available at the site itself, the additional expenditure of transporting soil cover is avoided. The analysis of soil available is necessary. Hydrogeological investigations are necessary as the rain water percolating through solid wastes tends to carry pollutants to the underlying strata, hence the pollution load contributed by such leachate will cause pollution of ground water. An impermeable barrier in the form of a puddle clay blanket should be provided to avoid leachate contamination. To avoid surface water pollution, the water course flowing across the site should be diverted and the surface water due to precipitation may be prevented from reaching the water course by an impermeable barrier.

For maintaining the site in proper working condition during monsoons, it is necessary to provide all-weather access roads to avoid slipperiness for mechanical equipment. Dewatering equipment is required in the trench method to remove the water that has filled the trenches. Firefighting equipment may be used to extinguish the fires at the site, which are caused due to hot ashes and combustible material. Airborne litter problems due to high paper content can be overcome by a movable screen of wire mesh. While the problem of dust could be overcome by sprinkling water over the deposited waste, proper drainage will avoid the excessive pounding of water on landfill sites, which may seep through. Rodents could be avoided by using either a proper covering material or rodent poison. The problem of birds is serious if the site is situated near airports and could be overcome by providing prompt covering. Flies and mosquitoes, if present, could be eliminated by using suitable insecticides. Due to anaerobic decomposition of organic matter in the waste, CH_4 and H_2S gases are produced. To avoid fire hazards due to such gases, suitable drains should be provided so that they may be safely let out or burnt. Due to the presence of decomposed organic material within the fill, the reclaimed land could be used for locating parks and playgrounds.

The cost includes the acquiring of site for landfilling, transportation charges, maintenance of civil works, and staff. The cost of the manual landfilling method works out to be Rs. 2/ to Rs. 8/ per tonne depending on the quantity of waste handled and soil available.

2.7.1.1 Design of Landfill

A landfill design life will comprise of an active period and a closure and postclosure period. The active period may typically range from 10 to 25 years depending on the availability of land area. The closure and postclosure period for which a landfill will be monitored and maintained will be 25 years after the active period is completed. The volume of waste to be placed in a landfill will be computed for the active period of the landfill taking into account (1) the current generation of water per annum, and (2) the anticipated increase in rate of waste generation on the basis of past records or population growth rate.

The required landfill capacity is significantly greater than the waste volume it accommodates. The actual capacity of the landfill will depend upon the volume occupied by the liner system and the cover material (daily, intermediate, and final cover), as well as the compacted density of the waste. In addition, the amount of settlement a waste will undergo due to overburden stress and due to biodegradation should also be taken into account. The density of waste varies on account of large variations in waste composition, degree of compaction, and state of decomposition. Densities may range as low as 0.40 t/cu.m. to 1.25 t/cu.m. For planning purposes, a density of 0.85 t/cu.m. may be adopted for biodegradable wastes with higher values and 1.1 t/cu.m. for inert waste.

Settlement of the completed waste mass beneath the final cover will inevitably occur as a result of the consolidation of waste within a landfill site. Initial settlement occurs predominantly because of the physical rearrangements of the waste material after it is first placed in the landfill. Later settlement mainly results from biodegradation of the waste, which in turn leads to further physical settlement. A typical allowance of 10% can be made when usable landfill capacity is computed (less than 5% for incinerated/inert waste). The total landfill area should be approximately 15% more than the area required for landfill to accommodate all infrastructure and support facilities as well as to allow the formation of a green belt around the landfill. There is no standard method for classifying landfills by their capacity. However the following nomenclature is often observed in literature:

Small size landfill:	less than 5 hectare area
Small size landfill:	5 to 20 hectare area
Large size landfill:	greater than 20 hectare area

Landfill heights are reported to vary from less than 5 m to well above 30 m.

2.7.1.2 Landfill Layout

A landfill site will comprise the area in which the waste will be filled as well as additional area for support facilities. Within the area to be filled, work may proceed in phases with only a part of the area under active operation. The following facilities must be located in the layout: (1) access roads; (2) equipment shelters; (3) weighing scales; (4) office space; (5) location of waste inspection and transfer station (if used); (6) temporary waste storage and/or disposal sites for special wastes; (7) areas to be used for waste processing (e.g., shredding); (8) demarcation of the landfill areas and areas for stockpiling cover material and liner material; (9) drainage facilities; (10) location of landfill gas management facilities; (11) location of leachate treatment facilities; and (12) location of monitoring wells.

2.7.1.3 Landfill Selection

Landfills may have different types of selections depending on the topography of the area. The landfills may take the following forms: (1) above-ground landfills (area landfills); (2) below-ground landfill (trench landfills); (3) slope landfills; (4) valley landfills (canyon landfills); and (5) a combination of the above.

Above-Ground Landfill (Area Landfill): The area landfill is used when the terrain is unsuitable for the excavation of trenches in which to place the solid waste. High ground-water conditions necessitate the use of area-type landfills. Site preparation includes the installation of a liner and leachate control system. Cover material must be hauled in by truck or earth-moving equipment from adjacent land or from borrow-pit areas.

Below-Ground Landfill (Trench Landfill): The trench method of landfilling is ideally suited to areas where an adequate depth of cover material is available at the site and where the water table is not near the surface. Typically, solid wastes are placed in trenches excavated in the soil. The soil excavated from the site is used for daily and final cover. The excavated trenches are lined with low-permeability liners to limit the movement of both landfill gases and leachate. Trenches vary from 100 to 300 m in length, 1 to 3 m in depth, and 5 to 15 m in width with side slopes of 2:1.

Slope Landfill: In hilly regions it is usually not possible to find flat ground for landfilling. Slope landfills and valley landfills have to adopted. In a slope landfill, waste is placed along the sides of an existing hill slope. Control of inflowing water from hillside slopes is a critical factor in the design of such landfills.

Valley Landfill: Depressions, low-lying areas, valleys, canyons, ravines, dry borrow pits, etc. have been used for landfills. The techniques to place and compact solid wastes in such landfills vary with the geometry of the site, the characteristics of the available cover material, the hydrology and geology of the site, the type of leachate and gas-control facilities to be used, and the access to the site. Control of surface drainage is often a critical factor in the development of canyon/depression sites.

It is recommended that the landfill selection be arrived at keeping in view the topography, depth to water table, and availability of daily cover material.

2.7.1.4 Operating Methodology

Before the main design of a landfill can be undertaken, it is important to develop the operating methodology. A landfill is operated in phases because it allows the progressive use of the landfill area, such that at any given time a part of the site may have a final cover, a part being actively filled, a part being prepared to receive waste, and a part undisturbed.

The term "phase" describes a sub-area of the landfill. A "phase" consists of cells, lifts, daily cover, intermediate cover, liner and leachate-collection facility, gas-control facility, and final cover over the sub-area. Each phase is typically designed for a period of 12 months. Phases are generally filled from the base to the final/intermediate cover and capped within this period, leaving a temporary unrestored sloping face. It is recommended that a phase plan be drawn as soon as the landfill layout and selection are finalized. It must be ensured that each phase reaches the final cover level at the end of its construction period and that is capped before the onset of monsoons. For very deep or high landfills, successive phases should move from the base to the top (rather than horizontally) to ensure early capping and less exposed plan area of "active" landfills.

The term *cell* is used to describe the volume of material placed in a landfill during one operating period, usually one day. A cell includes the solid waste deposited and the daily cover material surrounding it. Daily cover usually consists of 15–30 cm of native soil that is applied to the working faces of the landfill at the end of each operating period. The purposes of daily cover are to control the blowing of waste materials, to prevent rats, flies, and other disease vectors from entering or exiting the landfill, and to control the entry of water into the landfill during operation.

A lift is a complete layer of cells over the active area of the landfill. Each landfill phase is comprised of a series of lifts. Intermediate covers are placed at the end of each phase; these are thicker than daily covers, and 45 cm or more remain exposed till the next phase is placed over it. A bench (or terrace) is commonly used where the height of the landfill will exceed 5 m. The final lift includes the cover layer. The final cover layer is applied to the entire landfill surface of the phase after all landfilling operations are completed. The final cover usually consists of multiple layers designed to enhance surface drainage, intercept percolating water, and support surface vegetation.

2.7.1.5 Estimation of Leachate Quality and Quantity

Leachate is generated on account of the infiltration of water into landfills and its percolation through waste as well as by the squeezing of the waste due to self-weight. Thus, leachate can be defined as a liquid that is produced when water or another liquid comes in contact with solid waste. Leachate is a contaminated liquid that contains a number of dissolved and suspended materials.

The important factors that influence leachate quality include waste composition, elapsed time, temperature, moisture, and available oxygen. In general, leachate quality of the same waste type may be different in landfills located in different climatic regions. Landfill operational practices also influence leachate quality.

Data on leachate quality has not been published in India. However, studies conducted by the Indian Institute of Technology, Delhi, National Environmental Engineering Research Institute (NEERI), Nagpur, and some state pollution control boards have shown groundwater contamination potential beneath sanitary landfills. Data on characteristics of leachates reported by Bagchi (1994), Tchobanoglous et al. (1993), and Oweis and Khera (1990) is as given in Table 2.7.

Assessment of leachate quality at an early stage may be undertaken to identify whether the waste is hazardous, to choose a landfill design, design or gain access to a leachate treatment plant, and develop a list of chemicals for the ground-water monitoring program. To assess the leachate quality, toxicity characteristic leaching procedure (TCLP tests) are to be followed. Laboratory leachate tests on municipal solid waste do not yield very accurate results because of heterogeneity of the waste as well as difficulty in simulating time-dependent field conditions. Leachate samples from old landfills may give some indication regarding leachate quality; however, this too will depend on the age of the landfill.

For the design of municipal solid waste landfills having significant biodegradable material as well as mixed waste, leachate quality has been universally observed to be harmful to ground-water quality. Hence, all landfills will be designed with a liner system at the base. A landfill may not be provided with a liner if and only if the following conditions can be satisfied:

1. If the waste is predominantly construction material—type inert waste without any undesirable mixed components (such as paints, varnish, polish, etc.) and if laboratory tests (such as TCLP tests) conclusively prove that the leachate from such waste is within permissible limits; and
2. If the waste has some biodegradable material, it must be proven through both laboratory studies on fresh waste and field studies (in old dumps) that the leachate from such waste will not impact the ground water in all the phases of the landfill and has not impacted the

Table 2.7 Constituents of leachates from municipal solid waste landfills

Constituent		Range mg/L	
Type	**Parameter**	**Minimum**	**Maximum**
Physical	pH	3.7	8.9
	Turbidity	30 JTU	500 JTU
	Conductivity	480 mho/cm	72,500 mho/cm
Inorganic	Total suspended solids	2	170,900
	Total dissolved solids	725	55,000
	Chloride	2	11,375
	Sulfate	0	1850
	Hardness	300	225,000
	Alkalinity	0	20,350
	Total Kjeldahl nitrogen	2	3320
	Sodium	2	6010
	Potassium	0	3200
	Calcium	3	3000
	Magnesium	4	1500
	Lead	0	17.2
	Copper	0	9.0
	Arsenic	0	70.2
	Mercury	0	3.0
	Cyanide	0	6.0
Organic	COD	50	99,000
	TOC	0	45,000
	Acetone	170	110,000
	Benzene	2	410
	Toluene	2	1600
	Chloroform	2	1300
	1,2 dichloromethane	0	11,000
	Methyl ethyl ketene	110	28,000
	Naphthalene	4	19
	Phenol	10	28,800
	Vinyl Chloride	0	100
Biological	BOD	0	195,000
	Total coliform bacteria	0	100

Source: Range of constituents observed from different landfills. Table compiled from data reported by Bagchi (1994), Tchobanoglous et al. (1993) and Oweis and Khera (1990).

ground water or the subsoil so far in old dumps. Such a case may occur at sites where the base soil may be clay of permeability less than 10^{-7} cm/s for at least 5-m depth below the base and where water table is at least 20 m below the base. A leachate collection facility would have to be provided in all such cases.

The quantity of leachate generated in a landfill is strongly dependent on the quantity of infiltrating water. This in turn is dependent on weather and operational practices. The amount of rain falling on a landfill to a large extent controls the leachate quality generated. Precipitation depends on geographical location.

A significant quantity of leachate is produced from the active phases of a landfill under operation during the monsoon season. The leachate quantity from those portions of a landfill that have received a final cover is minimal.

Generation Rate in Active Areas: The leachate generation during the operational phase from an active area of a landfill may be estimated in a simplified manner as follows:

$$\begin{aligned}\text{Leachate volume} = {} & (\text{volume of precipitation}) \\ & + (\text{volume of poresqueeze liquid}) \\ & - (\text{volume lost through evaporation}) \\ & - (\text{volumeof water absorbed by the waste})\end{aligned}$$

Generation Rate After Closure: After the construction of the final cover, only that water which can infiltrate through the final cover percolates through the waste and generates leachate. The major quantity of precipitation will be converted to surface runoff and the quantity of leachate generation can be estimated as follows:

$$\begin{aligned}\text{Leachate volume} = {} & (\text{volume of precipitation}) \\ & - (\text{volume of surface runoff}) \\ & - (\text{volume lost through evapotranspiration}) \\ & - (\text{volume of water absorbed by waste and intermediate} \\ & \quad \text{soil covers})\end{aligned}$$

For landfills that do not receive runoff from outside areas, a very approximate estimate of leachate generation can be obtained by assuming it to be 25–50 percent of the precipitation from the active landfill area and 10–15 percent of the precipitation from covered areas. This is a thumb rule and can only be used for preliminary design.

For detailed design, computer-simulated models, e.g., hydraulic evaluation of landfill performance (HELP), have to be used for estimation of leachate quantity generation. It is recommended that for design of all major landfills, such studies be conducted to estimate the quantity of leachate.

2.7.1.6 Design and Construction of Landfill Liners

The liner system at the base and sides of a landfill is a critical component of the landfill that prevents ground-water contamination. Design and construction procedures of two elements of the liner system, the compacted clay/amended soil and the geomembrane, are discussed.

Compacted Clays and Amended Soils: The selection of material to be used in a soil-barrier layer will usually be governed by the availability of materials, either at site or locally in nearby areas. The hierarchy of options is as follows:

1. Natural clay will generally be used as a mineral component of a liner system where suitable clay is available onsite or nearby.
2. If clay is not available, but there are deposits of silts (or sands), then formation of good-quality bentonite-enhanced soil/amended soil may be economical.

Compacted Clays: Wherever suitable low-permeability natural clay materials are available, they provide the most economical lining material and are commonly used. The basic requirement of a compacted clay liner is that it should have permeability below a prespecified limit (10^{-7} cm/s) and that this should be maintained during the design life. Natural clay available in-situ is usually excavated and recompacted in an engineered manner. If clay is brought from nearby areas, it is spread in thin layers and compacted over the existing soil. The quality of the in-situ clay may be good enough to preclude the requirement of a compacted clay liner, only if it has no desiccation cracks and is homogeneous as well as uniformly dense in nature.

Amended Soils: When low-permeability clay is not available locally, in-situ soils may be mixed with medium-to-high plasticity imported clay, or commercial clays such as bentonite, to achieve the required low hydraulic conductivity. In terms of soil, bentonite admixtures are commonly used as low permeability amended soil liners. Generally, well-graded soils require 5–10 percent by dry weight of bentonite, while uniformly graded soils (such as fine sand), may typically require 10–15 percent bentonite. The most commonly used bentonite admixture is sodium bentonite. Calcium bentonite may also be used, but more bentonite may be needed to achieve the required permeability, because it is more permeable than sodium bentonite.

It is not necessary that the bentonite should be the only additive to be considered for selection. Medium-to-high plasticity clays from not too distant areas can also be imported and mixed with the local soils. Usually high quantities of clays (10–25 percent) are required to achieve the required permeability. Nevertheless, these may sometimes prove to be more economical than bentonite-amended soils and their permeabilities may not be significantly influenced by leachate quality. A competent barrier made of compacted soils, i.e., clays or amended soils, is normally expected to fulfill the following requirements:

1. hydraulic conductivity of 10^{-7} cm/s or less
2. thickness of 100 cm or more.
3. absence of shrinkage cracks due to desiccation
4. absence of clods in the compacted clay layer
5. adequate strength for stability of liner under compressive loads as well as alongside slopes
6. minimal influence of leachate on hydraulic conductivity

Clays of high plasticity with very low values of permeability (usually well below the prescribed limit), exhibit extensive shrinkage on drying and tend to form large clods during compaction in the relatively dry state. Their permeability can also increase on ingress of certain organic leachates. Well-compacted inorganic clays of medium plasticity, either natural or amended, appear to be most suitable for liner construction.

Soil with the following specifications would prove to be suitable for liner construction:

- percentage fines: between 40% and 50%
- plasticity index: between 10% and 30%
- liquid limit: between 25% and 30%
- clay content: between 18% and 25%

It is necessary to perform detailed laboratory tests and some field-trial tests prior to liner construction to establish that the requirements pertaining to permeability, strength, leachate compatibility, and shrinkage are met.

The design process for a compacted soil liner consists of the following steps:

1. Identification of borrow area or source of material (in situ or nearby);
2. For in situ soils, conducting field permeability tests to assess suitability of the natural soil in its in situ condition;
3. Laboratory studies on liner material (from in situ or nearby locations), comprising of soil classification tests, compaction tests, permeability tests, strength tests, shrinkage tests, and leachate compatibility tests;
4. Identification of source of additive, if natural soil does not satisfy liner requirements, i.e., natural clay from not too distant areas or commercially available clay such as bentonite;

5. Laboratory studies (as detailed in (3) above) on soil-additive mixes using different proportions of additive to find minimum additive content necessary to achieve the specified requirements;
6. Field trial on test pads, to finalize compaction parameters (layer thickness, number of passes, speed of compactor), as well as to verify that field permeability of the compacted soil lies within prespecified limits.

For amended soils, the following tests should be conducted to arrive at the minimum additive content.

Additive Composition: Grain-size distribution, plasticity tests, and mineralogy tests are performed to identify the clay content, activity, and clay mineralogy of the additive.

Host-Material Composition: Grain-size distribution and plasticity tests are performed on the host material, to assess that the host material will mix readily with the additive. Clean sands, silty sands, and nonplastic silts usually mix readily with clays and bentonites. Cohesive hosts are more difficult to mix due to the balling effect yielding uneven mixing. The host material must be sufficiently dry for proper mixing.

Soil-Additive Compaction Tests: Standard Proctor (or modified) tests are undertaken with variable quantities of additives mixed to the soil, usually in increments of 2–5 percent. The influence of the additive on dry density and optimum moisture content should be evaluated.

Soil-Additive Permeability Tests: Permeability tests are conducted on compacted-then-saturated samples of amended soil with different percentages of additive, where each sample is compacted to maximum density at optimum water content. The hydraulic conductivity usually decreases with increasing additive content. It is possible to identify a minimum additive content, from a series of tests, which may be required to achieve the desirable hydraulic conductivity.

Analysis of Laboratory Results: Field engineers usually require a compaction specification, which states the minimum acceptable dry density as well as the acceptable range of water content. It is usually possible to arrive at a narrow acceptable range of water content and dry density. A step-by-step process of elimination is to be adopted to identify this acceptable range of water content and dry density, which should then be communicated to the field engineer.

The construction of a field trial test pad prior to undertaking construction of the main liner has many advantages. One can experiment with compaction equipment, water content, number of passes of the equipment, lift thickness, and compactor speed. Most importantly, one can conduct extensive testing, including quality control testing and hydraulic conductivity tests, on the test pad. The test pad should have a width that is significantly more than the width of the construction vehicles (>10 m) and greater length. The pad should ideally be the same thickness as the full-sized liner, but may sometimes be thinner. The in situ hydraulic conductivity may be determined by the sealed double ring in filtrometer method. In in situ tests on test pads, the hydraulic conductivity is measured under zero overburden stress. Hydraulic conductivity decreases with increasing overburden stress. The hydraulic conductivity measured on a test pad, should be corrected for the effects of overburden stress, based on results of laboratory conductivity tests performed over a range of compressive stresses.

Compacted Clays: The sequence of construction for compacted clay liners is as follows:

1. Clearing of borrow area by removal of shrubs and other vegetative growth;
2. Adjustment of water content in the borrow area, sprinkling or irrigating for increasing the water content and ripping and aerating for lowering the water content;
3. Excavation of material;
4. Transportation to site in haulers or through conveyor systems (short distance);
5. Spreading and leveling of a thin layer (lift) of soil (of thickness about 25 cm);
6. Spraying and mixing water for final water-content adjustment;
7. Compaction using rollers;
8. Construction quality assurance testing;
9. Placement of next lift and repetition of process till final thickness is achieved.

The twofold objectives of soil compaction are (1) to break and remold the clods into a homogeneous mass, and (2) to densify the soil. If the compaction is performed such that the required density at the specified moisture content is obtained, the required permeability will be achieved in the field. Regulations generally require that a minimum 100-cm thick compacted clay liner be constructed. This thickness is considered necessary so that any local imperfections during construction will not cause hydraulic short-circuiting of the entire layer. Compacted soil liners are constructed in a series of thin lifts. This allows proper compaction and homogeneous bonding between lifts. Generally, the lift thickness of clay liners is 25 to 30 cm before compaction and about 15 cm after compaction. It is important that each lift of clay liner be properly bonded to the underlying and overlying lifts. If this is not done, a distinct lift interface will develop, which may provide hydraulic connection between lifts.

Sheep-foot rollers are best suited for compacting clay liners. Rollers with fully penetrating feet have a shaft about 25 cm long. Unlike partially penetrating rollers (pad-footed rollers), the fully penetrating sheep-foot roller can push through an entire soil lift and remold it. In addition to increasing bonding between lifts, one should maximize the compactive energy by considering factors such as roller weight, area of each foot, number of passes, and the speed of the roller.

The lifts are typically placed in horizontal layers. However, when liners are constructed on the side slopes, the lifts can be placed either parallel to the slope (for slopes up to 2.5 horizontal, 1 vertical, due to limitations of compaction equipment) or in horizontal lifts. Horizontal lifts require a width that is at least the width of one piece of construction equipment (usually 3–4 m).

Amended Soils: The process of construction of amended soil liners is similar to that for compacted clay liners with the modification that the additive is introduced into the soil after the excavation stage. Additives such as bentonite can be introduced in two ways, namely by in-place mixing or by central plant method. In the latter technique, the soil and additive are mixed in a pug mill or a central mixing plant. Water can also be added in the pug mill either concurrently with bentonite or in a separate processing step. The central mixing plant method is more effective than in-place mixing and should be adopted. The use of small truck-mounted concrete-batching plants for mixing bentonite can also be examined.

The quality of the mix must be checked to ensure uniformity and correctness of the additive. A minimum of five trial runs should be made to check the quality of the mix visually and using grain-size analysis. The permeability should also be checked using the field mix, compacted in the laboratory.

During construction, quality control is exercised to ensure that the constructed facility meets the design specifications. Borrow-area material control and amended soil control involves the following tests: (1) grain size distribution, (2) moisture content, (3) Aterberg's limits, (4) laboratory compaction tests, and (5) laboratory permeability tests. The frequency of testing varies from one test per 1000 cu.m, to one test per 5000 cu.m.

Compacted soil-liner control involves the following tests: (1) in situ density measurements, (2) in situ moisture content measurements, (3) laboratory permeability tests on undisturbed samples, (4) in situ permeability tests, (5) grain-size distribution, and Atterberg's limits of compacted samples. The frequency of testing for in situ density and moisture content may be as high as 10 tests/hectare/lift whereas the other tests may be conducted at a lower frequency of about 2 tests/hectare/lift.

The geomembranes with thickness of 1.5 mm are to be laid over the compacted clay/amended soil with no gaps along the surface of contact. The geomembrane is normally expected to meet the following requirements:

1. It should be impervious.
2. It should have adequate strength to withstand subgrade deformations and construction loads.
3. It should have adequate durability and longevity to withstand environmental loads.
4. The joints/seams must perform as well as the original material.

The specifications for geomembrane liners are presented in the Table 2.8. The specifications are only suggestive and need to be refined by a geosynthetics specialist for each landfill site shown in Fig. 2.1.

The components that have to be designed/checked for in the case of geomembranes are anchor trench, sliding along slopes, allowable weight of vehicle, uneven settlement, and panel-layout plan.

Although the construction activities for geomembrane installation are not as time consuming as clay liner construction, the quality control tests are intensive. The surface of compacted clay/amended soil must be properly prepared for installation of synthetic membrane. The surface must not contain any particles greater than 1.25 cm (0.5 in.) size. Larger particles may cause protuberance in the liner. The panel layout plan should be made in advance so that travel of heavy equipment on the liner can be avoided. In no case should it be allowed on the liner. Seaming of panels within 1.0 m of the leachate collection line location should be avoided if possible; this issue can be finalized during the layout plan. The subbase must be checked for footprints or similar depressions before laying the liner. The crew should be instructed to carry only the necessary tools and not to wear any heavy boots (tennis shoes are preferred). Laying of the synthetic membrane should be avoided during high winds (24 kmph or more). Seaming should be done within the temperature range specified by the manufacturer.

Table 2.8 Values for geomembranes measured in performance tests

S. no.	Property	Typical value
1.	a. Thickness	1.5 mm
	b. Density	0.94 gm/cc
2.	Roll width × length	6.5 m × 150 m
3.	Tensile strength	
	a. Tensile strength at yield	24 kN/m
	b. Tensile strength at break	42 kN/m
	c. Elongation at yield	15%
	d. Elongation at break	700%
	e. Secant modulus (1%)	500 MPa
4.	Toughness	
	a. Tear resistance (initiation)	200 N
	b. Puncture resistance	480 N
	c. Low-temperature brittleness	94°F
5.	Durability	
	a. Carbon black	2% Negligible strength
	b. Carbon black dispersion	changes after 1 month
	c. Accelerated heat aging	at 110°C A-1
S. no.	**Property**	**Typical value**
6	Chemical resistance	
	a. Resistance to chemical waste mixture	10% strength change over 120 days
	b. Resistance to pure chemical reagents	10% strength change over 7 days
7	Environmental stress-crack resistance	1500 h
8	Dimensional stability	+2%
9	Seam strength	80% or more (of tensile strength)

Figure 2.1 Landfill site.

Several types of seaming methods are available. The following are some of the most commonly used seaming techniques: thermal hot air, hot wedge fusion, extrusion welding (fillet or lap), and solvent adhesive. The manufacturer usually specifies the types of seaming to be used and in most cases provides the seaming machine. Manufacturer's specifications and guidelines for seaming must be followed. Seaming is more of an art even with the automatic machines. Only persons who are conversant with the machine and have some actual experience should be allowed to seam. For HDPE, hot wedge fusion and extrusion welding–type seaming are commonly practiced. Geomembranes must be covered with protective soil as soon as possible. Enough volume of soil should be stockpiled near the site so that it can be spread on the finished membrane as soon as the quality control test results are available and the final inspection is over. Synthetic membranes can be damaged by hoofed animals. Bare membrane should be guarded against such damage by fencing the area or by other appropriate methods.

At least 30 cm of fine sand or silt or similar soil should be spread on the membrane as a protective layer. The soil should be screened to ensure that the maximum particle size is 1.25 cm or less. The traffic routing plan must be carefully made so that the vehicle(s) does not travel on the membrane directly. Soil should be pushed gently by a light dozer to make a path. Dumping of soil on the membrane should be avoided as much as possible. One or two main routes with extra thickness of soil (60–90 cm) should be created for use by heavier equipment for the purposes of soil moving. Even the utmost precaution and quality control during installation will be meaningless if proper care is not taken when covering the membrane. Slow and careful operations are the key to satisfactory soil spreading.

The geomembrane bid specification should include warranty coverage for transportation, installation, and quality control tests. The cost of a project may increase due to the warranty. The experience of the company (both in manufacturing and installation), quality control during manufacturing and installation, and physical installation should be asked about in the bid so proper comparisons among different bidders can be made.

Quality Control Before and During Geomembrane Installation: Tests of several physical properties of the membrane must be performed before installation. Usually most of these tests are performed at the time of manufacturing in the manufacturer's laboratory. The owner may arrange for an independent observer to oversee the tests, conduct the tests in an independent laboratory, or use a split-sampling technique. This issue of responsibility for preinstallation quality control tests must be clearly mentioned or resolved during the biding process. The following are tests used for quality control purposes: (1) sheet thickness, (2) melt index, (3) percentage carbon black, (4) puncture resistance, (5) tear resistance, (6) dimensional stability, (7) density, (8) low temperature brittleness, (9) peel adhesion, and (10) bonded seam strength.

The quality control tests that are performed during installation include the following:

1. inspection of surface of compacted clay/amended soil layer
2. verification of the proposed layout plan
3. checking roll overlap
4. checking anchoring trench and sump
5. testing of all factory and field seams using proper techniques over full length
6. destructive seam strength test
7. patching up repair

A drainage layer is constructed over the protective soil layer placed on a geomembrane. It must have permeability greater than 10^{-2} cm/s. The 0.074 mm or less fraction content of the drainage blanket material should not be more than 5%. A clean coarse sand is the preferred material for the drainage blanket, however, gravel may also be used for this purpose. When a layer of gravel is used as a drainage blanket; the fines from the waste may migrate and clog the blanket. A filtering medium design approach may be used in designing a graded filter over a gravel drainage blanket.

The quality control tests include tests for grain-size analysis and permeability. Usually one grain-size analysis for each 1000 cu.m and one permeability test for each 2000 cu.m of material used is sufficient. For smaller volumes a minimum of four samples should be tested for each of the above properties. The permeability of the material should be tested at 90% relative density. Sand blankets will be placed in leachate collection trenches as specified by the designer of leachate collection pipes.

2.7.2 Composting

Composting is a process in which waste is decomposed and converted into manure. This can be achieved microbiologically and with vermin.

2.7.2.1 Microbiological Composting

Composting is a process of decomposition and consists of the successive overlapping of microbiological processes, the intensity of which is shown by the production of heat and gaseous products of metabolism. The rapidly changing environmental conditions first result in an explosion-like multiplication of numbers and species of microorganisms, followed by a rapid pasteurization, which is necessary for the destruction of pathogenic species.

The role of mold fungi in composting is very significant and a study was made by Dr. G. Parkas of Giessen University. The study showed that mold fungi play an important role in the decomposition of organic matter. The molds are distributed everywhere in nature and are far less inhibited by unfavorable living conditions, such as low moisture content, low temperature, and high acidity of the substrate than are the bacteria. Further, it was noticed that there was more intensive proliferation of fungi at O_2-dominant areas in a test pile than in the interior of the pile where there is less oxygen. The above observation holds true for both mesophile and thermopile species of fungi.

The relation of fungi to the moisture content of the piles was also confirmed, with more fungi being present in dry regions than in wetter regions. Also with the increase of temperature mesophiles multiplied rapidly at about 37°C and gradually subsided on further increase in temperature and stopped at 67°C. Upon cooling, the mesosphelic species again appeared and were more numerous in dry areas. The thermophiles, on the contrary, did not multiply initially and their numbers decreased with increased temperature and vanished at 67°C after 3 days. Later they multiplied intensively during the decrease in temperature. Fungi developed to the greatest extent under optimum conditions in the most prominent zones of the piles with white and pigmented mycelia.

Further, it was noted that the initial increase in fungi does not originate from uniform growth of the species present in the raw refuse but from a rapidly developing change in microbial flora. Fungi present in raw refuse were uniformly of a few species, up to 70% consisting of *Gotrichium*. All the species disappeared above 87°C. Further, while cooling, the fungi flora appeared again with new species.

Gardeners, horticulturists, and wine growers prefer compost; farmers on the contrary, prefer as cheap a product as possible. With a favorable long-term effect it has been shown that delayed benefits of raw refuse are superior to those of the compost. Hence, it is thought that raw urban refuse, having few pathogens, could be used directly on the soil after the removal of objectionable materials. But care should be taken to see that the composted product is hygienically safe especially when the compost consists of sewage sludge, as that contains pathogenic microorganisms. Therefore, destruction of pathogens becomes necessary.

Experiments have shown that pathogens were filled in a raw refuse sludge mixture of initial 40–60% moisture content after about 14 days of action at a temperature of 55–60°C with at least one turning. Further compost files of relatively high moisture content developed anaerobic zones that were favorable for pathogens. This can be avoided by converting anaerobic zones to aerobic zones through turning.

Further, the use of compost containing sludge mixtures for agriculture involving animals demands complete destruction of pathogen-causing diseases such as hog, cholera, chicken plague, anthrax, etc. It has also been shown that temperature is not the sole criteria for the destruction of pathogens. Certain antibiotic agents develop during composting which exert an antagonistic action on the pathogens. Hence, it can be concluded that the composting of raw refuse with sludge mixtures will result in hygienically safe compost that can be used for agricultural purposes where animals used in the process. Compost is beneficial for crop production for the following reasons:

1. Compost prepared from municipal refuse contains about 1% each of NPK.
2. During composting, the plant nutrients are converted to such forms that get released gradually over a longer period and do not get leached away easily.
3. It is known to contain trace elements such as Mn, Cu, Bo, Mo, which are essential to the growth of plants.
4. It is a good soil conditioner and increases the texture of soil, particularly in light sandy soil.
5. It improves the ion exchange and water-retaining capacity of the soil.
6. The organic matter in soil in tropical climates gets depleted rapidly by microbial activity. Compost adds stabilized organic matter, thus improving the soil.
7. It increases the buffering capacity of the soil.

Hence, compost application to soil is beneficial, but compost cannot be an alternate to chemical fertilizer; each of them has a specific role to play. It is desirable that compost is used in conjunction with chemical fertilizers to obtain optimum benefits. It has been successfully demonstrated that best results are obtained when the two are used together in certain proportions. The yield in such cases has been reported to be much higher than that obtained when they are used separately.

The organic material present in the municipal wastes can be converted to a stable form either aerobically or anaerobically. During aerobic decomposition, aerobic microorganisms oxidize organic compounds into CO_2, NO_2, and NO_3. Carbon from organic compounds is used as a source of energy, while nitrogen is recycled. Due to the exothermic reaction, temperature of the mass rises. Anaerobic microorganisms, while metabolizing nutrients, break down the organic compounds by a process of reduction. A very small amount of energy is released during the process, and temperature of the composting mass does not rise much. The gases evolved are mainly CH_4 and CO_2. As anaerobic decomposition of organic matter is a reduction process, the final product is subject to some minor oxidation when applied to land.

Factors Affecting Composting Processes: The factors affecting the composting process are organisms, use of cultures, moisture, temperature, and C/N ratio.

Organisms: Aerobic composting is a dynamic system in which bacteria, actinomycates, fungi and other biological forms are actively involved. The relative predominance of one species over another depends upon the constantly changing available food supply, temperature, and substrate conditions. In this process, facultative and obligate aerobic forms of bacteria, actinomycetes, and fungi are most active. Mesosphilic forms are predominant in the initial stages, which soon give way to thermophilic bacteria and fungi. Except in the final stages of composting when the temperature drops, actionomycetes and fungi are confined to 5–15 cu.m. of the outer surface layer. If the turning is not carried out frequently, increased growth of actinomycetes and fungi in the outer layers imparts a typical greyish-white color. Thermophilic actionomycetes and fungi are known to grow well in the range of 45–60°C.

Attempts have yet to be made to identify the role of different organisms in the breakdown of different materials. Thermophilic bacteria are mainly responsible for the breakdown of proteins and other readily biodegradable organic matter. Fungi and actinomycetes play an important role in the decomposition of cellulose and lignins. Among the actiomycetes, *Stroptomyces sp.* are common in compost, the latter being more prevalent. The common fungi in compost are *Penicillium dupontii* and *Aspergillus fumigates.*

During the development of composting systems various innovators have come forward with inoculums, enzymes, etc., which are claimed to hasten the composting process. Investigations carried out by a number of workers have shown that they are unnecessary. When the environmental conditions are appropriate, indigenous bacteria, better adapted to municipal refuse than forms attenuated under laboratory conditions, rapidly multiply and carry out necessary decomposition. Since the process is dynamic and as any specific organisms can survive over a specific environmental range, as one group starts diminishing, another group of organisms starts flourishing. Thus, in such a mixed system, bacteria develop and multiply to the addition of similar and extraneous organisms such as an inoculum is superfluous. Such inoculums may, however, be important while composting some industrial and agricultural solid wastes that do not have the required indigenous bacterial population.

Moisture: Moisture replaces air from the interspace between particles. Too low a moisture content reduces the metabolic activity of organisms, whereas anaerobic conditions would set in if the moisture content is too high. It has been shown that the optimum moisture content for composting is in the range of 50–60 percent. Moisture required for satisfactory aerobic composting will depend on the materials used. High moisture content will be required if straw and strong fibrous materials are present to soften the fibers. Moisture content higher than 50–60 percent can be used in mechanically aerated digesters. In anaerobic composting, the moisture required will depend upon the method of storage and handling.

Temperature: During anaerobic decomposition, 26 kcal is released per gram mole of glucose as against 484–674 kcal under aerobic conditions. As refuse has good insulation properties, the heat of the exothermic biological reaction accumulates resulting in an increase of temperature of the decomposing mass. Loss of heat will occur from the surface and hence the larger the exposed surface area per unit weight of the composting mass, the larger will be the heat loss. When windrows are turned heat loss occurs, resulting in a drop in temperature, but it rises during active decomposition to as high as 70°C. Addition of water to the composting mass results in a drop in temperature. The temperature will tend to drop only when the conditions become anaerobic or the active period of decomposition is over. During anaerobic composting, a small amount of heat is released, much of which escapes by diffusion, conduction, etc. Thus, the temperature rise will not be appreciable. The Table 2.9 gives

Table 2.9 Temperature and time of exposure needed for destruction of some common parasites and pathogens

S. no.	Organisms	Time and temperature
1.	*Salmonella typhosa*	No growth beyond 46°C, death in 30 min at 55–60°C and 20 min at 60°C, destroyed in a short time in compost environment
2.	*Salmonella sp.*	In 1 h at 55°C and in 15–20 min at 60°C
3.	*Shigella sp.*	In 1 h at 55°C
4.	*Escherichia coli*	In 1 h at 55°C and in 15–20 min at 60°C
5.	*Entamoeba histolytica cysts*	In few minutes at 45°C and in a few seconds at 55°C
6.	*Taenia saginata*	In few minutes at 55°C
7.	*Trichinella spiralis larvae*	Quickly killed at 55°C, instantly at 60°C
8.	*Brucella abortus* or *Brucella suis*	In 3 min at 62–63°C and in 1 h at 55°C
9.	*Micrococcus pyogenes var. aureus*	In 10 min at 50°C
10.	*Streptococcus pyogenes*	In 10 min at 54°C
11.	*Mycobacterium tuberculosis var. hominis*	In 15–20 min at 66°C or after momentary heating at 67°C
12.	*Corynenbacterai*	In 45 min at 55°C
13.	*Necator americanus*	In 50 min at 45°C
14.	*Ascaris lumbricoides* eggs	In 1 h at 50°C

the temperature and time of exposure for the destruction of some common pathogens and parasites.

During aerobic composting, when material is turned twice in 12 days, *Entamoeba hystolytica* gets killed, and when turned thrice in 36 days eggs of *Ascaris lumbricoides* are also destroyed. In an anaerobic process, the destruction of parasites and pathogens occurs due to long detention time in an unsuitable environment, biological antagonism, and natural die away. The destruction of pathogens and parasites cannot be assured in anaerobic processes. It has been seen that activity of cellulose enzymes gets reduced above 70°C and the optimum temperature range for nitrification lies in the range from 30° to 50°C, above which a large N_2 loss occurs. In the temperature range of 50–60°C, high nitrification and cellulose degradation occur and destruction of pathogens and parasites is also ensured.

C/N Ratio: Progress of decomposition in a composting mass is greatly influenced by C/N value. Since living organisms utilize about 30 parts of carbon for each part of nitrogen, an initial C/N of 30 would be most favorable for rapid composting. Research workers have reported optimum values ranging between 26 and 31 depending upon other conditions. C/N bring the ratio of available carbon to available nitrogen; some forms resistant to biological attack may not be readily available. In cases where C/N is not at desirable levels, straw, sawdust, paper, etc., are materials that can be used as carbon sources while blood, sludges, and slaughterhouse waste serve as good sources of nitrogen. The municipal waste in developed countries has C/N values up to 80, to which sewage sludge (C/N of 5 to 8) is added to keep the C/N ratio of the mixture (Table 2.10).

This partly solves the problem of treatment and disposal of sewage that would otherwise require costly methods such as vacuum filters, filter press, etc. Municipal refuse in India and in other developing countries has an initial C/N ratio of about 30, which does not need blending except in marginal cases. When initial C/N value is low, loss of nitrogen in the form of ammonia occurs, thus a large part of the nitrogen will get lost. Addition of sewage and sewage sludges will involve problems of smell and odor, handling, and transportation costs. Even when sewage is used as a source of moisture in composting, the bulk of the sewage will have to be treated otherwise. In view of this, addition of sewage sludge is not suitable in developing countries.

Aeration by natural process occurs in the superficial layers of the composting mass, while the inner layers tend to progressively turn anaerobic as the rate of oxygen replenishment cannot keep pace with utilization. It is hence necessary to bring the inner layers in contact with oxygen, which is accomplished by aeration by turning the material or by supplying compressed air. In temperate regions, the composting mass is enclosed and air is supplied at the rate of 1–2 m^3 of air/day/kg. In tropical regions where ambient temperatures are sufficiently high, composting is carried out in windrows that are turned periodically. Aerobic conditions can be

Table 2.10 Nitrogen conservation in relation to C/N ratio

Initial C/N ratio	Final % of nitrogen (dry weight basis)	% N_2 loss
20	1.44	38.8
20.5	1.04	48.1
22	1.63	14.8
30	1.21	0.5
35	1.32	0.5
76	0.86	—

maintained when the windrows are turned on alternate days. NEERI studies show that windrows of raw refuse remain aerobic (due to large void spaces) if turned after 5 days. The longer turning interval helps to reduce the cost.

2.7.2.2 Arrangement of Material During Turning of Windrows

Control of the Composting Process: The composting process needs to be regulated so as to ensure aerobic conditions and to stop when completed. If the process is not regulated properly the final C/N may be either too low or too high. If the C/N ratio of the final product is high, the excess carbon tends to utilize nitrogen in the soil to the build-up cell protoplasm resulting in "robbing" of nitrogen in the soil. When the final C/N ratio is too low, the product does not help improve the structure of the soil. When night soil or sludge is added to the composting mass the parasites and pathogens may survive in the final product if high temperatures are not maintained for required period. Temperature and stability tests should be used together to test the stage and completion of the process.

Composting Systems: The composting systems can be broadly grouped as (1) aerobic and (2) anaerobic. During the initial period of development of mechanical compost plants, a combination of anaerobic and aerobic methods were used (Beccari method). Composting in pits used an anaerobic process (Bangalore method). Aerobic systems can be operated either manually or mechanically in open windrows, pits, or in enclosed digesters. An open-windrow system is preferred in tropical regions while in temperate regions a closed-digester system is used. The pit method of aerobic composting is also known as the Indore method.

Indore and Bangalore Methods of Composting: In the Bangalore method, a layer of coarse refuse is first put at the bottom of a pit to a depth of 5–6 cm, which is 7.5 cm deeper for a 25-cm width at the pit edges. Night soil is poured to a thickness of 5 cm in the depressed portion and the elevated edges prevent its draining to the sides. On top of this, a second layer of refuse is spread, which sandwiches the night soil; this is repeated till it reaches a height of 30 cm above the edge of the pit. The top layer of refuse should be at least 25–30 cm thick. The top of the mass is rounded to avoid rain water entering the pit. Sometimes a top layer of soil is given to prevent fly-breeding. It is allowed to decompose for 4–6 months, after which the compost can be taken out for use.

The Indore method of composting in pits is similar to the above except that it is turned at specific intervals to help maintain aerobic conditions, which will ensure high temperature, uniform decomposition, as well as absence of flies and odor. While filling with refuse and night soil, about 60 cm on the longitudinal side of the pit is kept vacant for starting the turning operations. The first turning is manually carried out after 4–7 days using long-handled rakes and the second turning after 5–10 more days.

Further turning is not necessary and composting will be complete in a period of 13–27 days. Aerobic composting of refuse and night soil in windrows can also be carried out using windrows of more or less the same dimensions as the pit. The windrow method of aerobic composting is more popular for composting municipal refuse without night soil.

Mechanical Methods: Though manual methods are popular in India due to high labor cost and limitations of space, mechanical processes are preferred in industrialized countries. In 1922, Becceri in Italy patented a process using a combination of aerobic and anaerobic decomposition in enclosed containers. The first full-scale plant was established in 1932 in the Netherlands by a nonprofit utility company, VAM, using the Van Maanen process, in which raw refuse is composted in large windrows that are turned at intervals by mobile cranes moving on rails. The Dano process appeared in Denmark in 1930 and the Fdrazer Eweson process in the United States in 1969. Several patented processes have since been developed using different methods of preparation of refuse or digestion. A mechanical compost plant (Fig. 2.2) is a combination of various unit operations meant to perform specific functions.

Unit Operations in Mechanical Composting Plants: Refuse collected from the feeder area of the city reaches the plant site at a variable rate depending upon the distance of collection points. The compost plant, however, has to operate at a uniform input rate. It is hence necessary to have a balancing storage to absorb the fluctuations in refuse input, for which a storage hopper ranges from 8 to 24 hours' storage. The exact capacity will depend upon the schedule of incoming trucks, the number of shifts, and the number of days per week the plant and refuse-cleaning system work.

The refuse is then fed to a slowly moving (5 m/mt) conveyor belt and the nondecomposable materials such as plastics, paper, and glass are manually removed by laborers standing on either side of the conveyor belt. The thickness of material on the belt is kept below 15 cm to enable hand-picking by laborers provided with gloves and other protective equipment. The removed materials are stored separately so that they can (if possible) be commercially exploited. The metals are removed by either

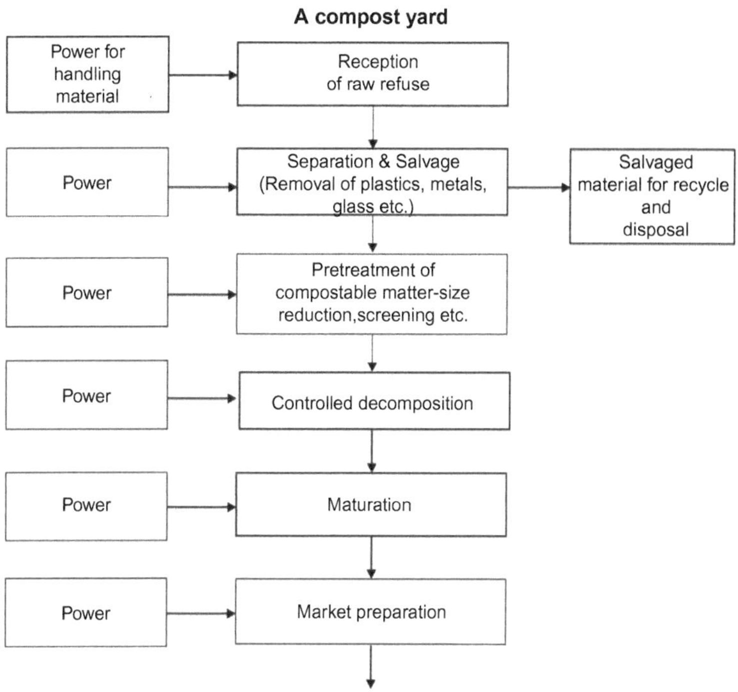

Figure 2.2 Flow chart for a mechanical compost plant.

suspended magnet system or a magnetic pulley system. In Indian refuse the metal content is low as most of it is reclaimed at the source itself. Metal remaining is either fine-sized or in an irrecoverable form. Magnetic removal is not efficient for low metal-content waste and hence not used in India.

Glass and metals are present in large proportions in the wastes from developed countries, for which ballistic separators are used. The materials are thrown with a force to take different trajectories depending on the density and get separated, but the operation is energy intensive. Glass and metals embedded in organic matter cannot be separated, making the unit ineffective.

The material after removal of most of the noncompostable material is subject to size reduction when the surface area per unit weight is increased for faster biological decomposition.

1. Hummer mills work at 600–1200 rpm and reduce the particle size by repeated hammer blows. These units are compact but consume more energy.

2. Rasp mills shear the material between a rotating arm moving at 4–6 rpm and a bottom plate with protruding pins. The units are relatively large (about 7 m in diameter and 7 m high for 12 tonne/hr capacity) and heavy, but consume less energy per unit weight of material. The capital cost of a hammer mill is less but the operating cost is more than a rasp mill. Some explosions have been reported in hammer mills due to at the presence of aerosol cans in the waste.

The material is now subjected to controlled decomposition after adjusting the moisture content to 50–60% by spraying water. The composting is carried out in:

1. *Closed Containers with Forced-Air Supply*: The container may be stationary (Earp Thomas) or may have a rotary motion (Dano System). Moisture, temperature, and air supply are continuously monitored and controlled. Enclosing the composting mass in containers is adopted in temperate region due to low ambient temperature and to protect it from rain and snow. These conditions lower the temperature of the decomposing mass and thus reduce the rate of reaction also.
2. *Windrow Composting*: In tropical regions with higher ambient temperature, composting in open windrows is to be preferred. The windrows have to be turned at suitable intervals to maintain the aerobic reactions. Compressed air supply will not be required in tropical regions. Turning of windrows can be carried out employing
   a. manual labor using buckets and shovels for smaller plants;
   b. front-end loaders having a bucket of 0.5–0.7 m^3 capacity and a 50-HP engine as in the case of earthmoving equipment. As the refuse is a light material its high-power requirement will prove uneconomical;
   c. clamshell bucket that will lift the material, move over a gantry girder and then drop it at another location. Elaborate structural fabrication of the yard will be required adding to the capital cost;
   d. mobile jib cranes are fixed in position and rotated in a horizontal plan but will add to the cost of number of such cranes will be required;
   e. augers moving in opposite directions mounted in pairs on a suitable frame moving horizontally on a pneumatic tires. The horizontal and rotary motion is provided with a suitable mechanism;
   f. machine with a rotating drum in the front to lift the material and pass it over a slant conveyor to the rear end of the machine.

A beater mechanism breaks the lumps in the material before it thrown out to reform the windrows (Fig. 2.3).

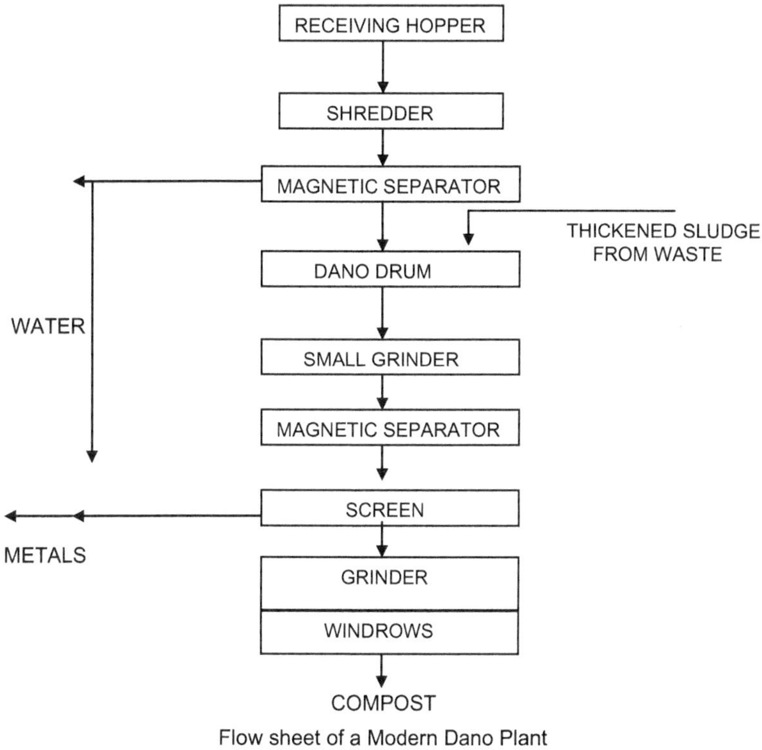

Figure 2.3 Flow sheet of a modern Dano plant.

The compost processed up to this stage is known as green or fresh compost, wherein the cellulose might not have been fully stabilized. The material is stored in large-sized windrows for a further period of 1–2 months. Compost will be used by farmers two or three times a year depending upon the cropping pattern. At the end of the storage period, the material is known as ripe compost; the ripe compost may further be processed for size reduction to suit kitchen gardens and horticultural requirements in urban areas.

Cost–Benefit Analysis: The degree of mechanization to be adopted will depend upon industrial and economic development, costs of labor and energy, and sociocultural attitudes of the community. A judicious combination of manual and mechanical methods will be required with due concern for public health aspects of the community as well as the workers and use of the product and acceptability by the farmers. A higher degree of mechanization will demand high energy inputs, which should be kept to a minimum. Reuse and recycling of valuable and available materials in the waste will recover part of the cost of production but not all of it. Public health protection and an esthetically clean environment are required, for which the community has to bear the cost. Compost production from community waste will need an element of social sharing of the cost (Fig. 2.4).

It is difficult to compare the costs for two different locations due to variation in size, plant components, method of operation, labor, energy, and land costs, as well as final disposal method, possibly due to lesser organic content in city refuse in industrialized countries. In Indian cities with more than 3,000,000 population, the capital cost varies from US $0.33 to 1.33 million (Rs. 3 million to 123 million) depending upon the degree of mechanization involved. The production cost of compost varies from US$2.65 to 8.88 (Rs. 25 to Rs. 80) per tonne.

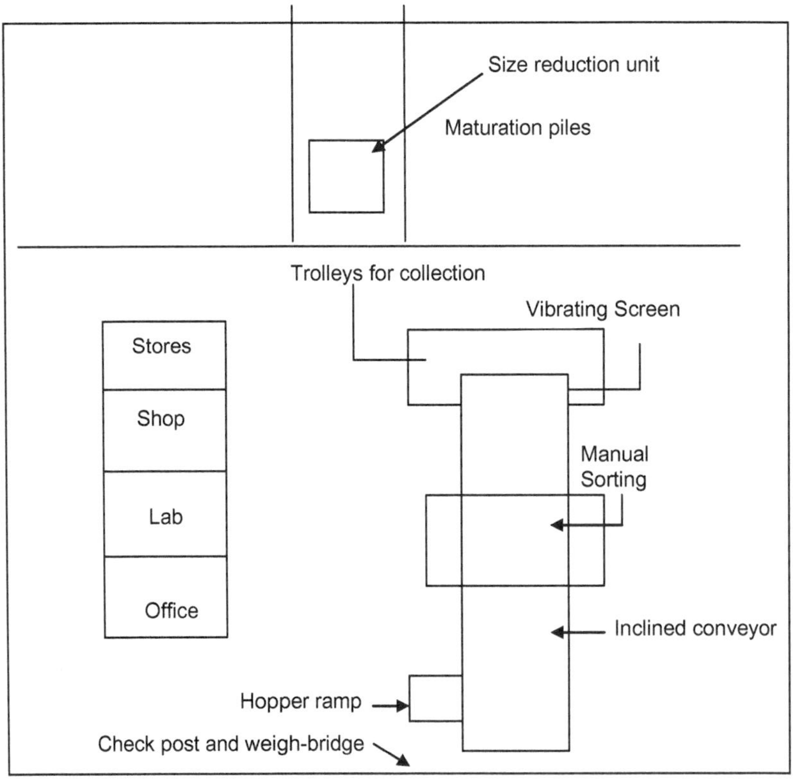

Figure 2.4 Flow sheet of Jaipur plant.

2.7.2.3 Vermicomposting

Vermiculture means the artificial rearing or cultivation of worms (earthworms) or the management of worms. Vermicomposting is a technology for turning organic waste into manure. Vermicompost is the excreta of the earthworm, which is rich in humus.

Vermicomposting (Fig. 2.5) is the process of degradation of organic wastes by earthworms to achieve three objectives, i.e., (1) to upgrade the value of organic waste materials so that they can be reused, (2) to produce upgraded materials in situ, and (3) to obtain a final product free of chemical or biological pollutants. The character of earthworm suited for vermicomposting are capable of inhabiting organic material in high percentage; are adaptable with respect to environmental factors; have a low incubation period and the smallest period of interval from hatching to maturity; have a high growth rate, consumption, digestion, and assimilation rates; and have the least vermin-stabilization time (period of inactivity after initial (Fig. 2.6) inoculation of organic wastes). The types of earthworms popularly used for vermin composting are *Epigamic earthworm, Eudrilus eugeniae, Eisenia foetida and Perionyx excavates.* The source of waste generation and composting materials are in Table 2.11.

Methods of Vermicomposting: There are two methods having practical application:

1. Solid waste materials are spread out over the soil surface, for incorporation directly into the soil by earthworms for burial and decomposition.
2. Wastes are stacked into heaps or placed in bins, where they are treated like compost heaps and earthworms are released.

Figure 2.5 Vermicomposting.

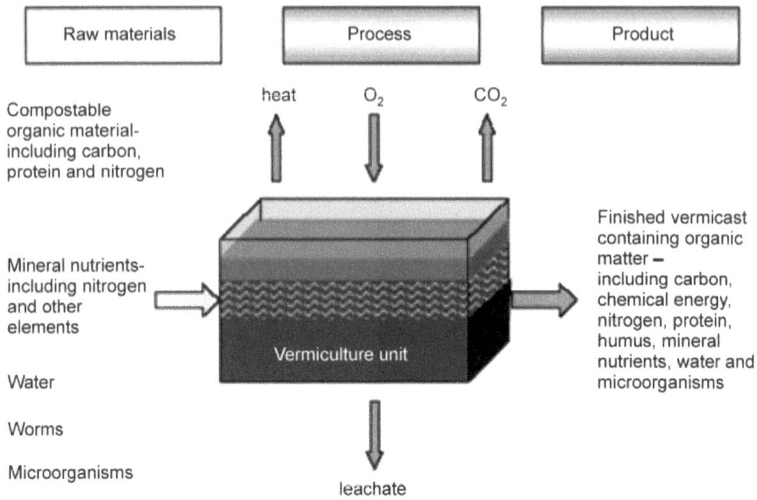

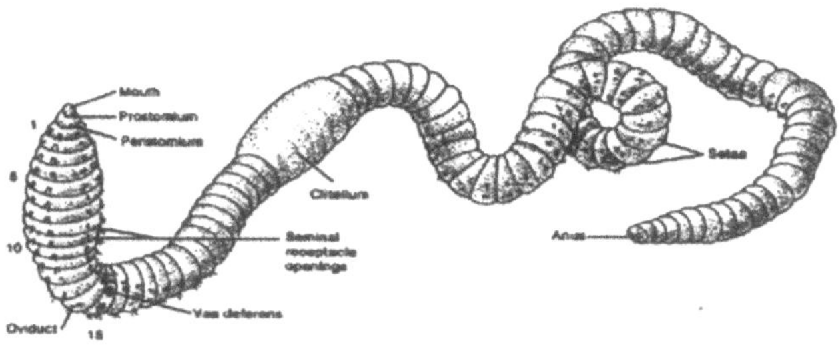

Figure 2.6 Important parts of earthworm.

Table 2.11 Source of waste generation and composting materials

Source of waste generation	Utilizable waste for vermicomposting
Agriculture waste	
1. Agricultural fields	Stubble, weeds, husk, straw, and farmyard manure
2. Plantations	Stems, leaf matter, fruit rind, pulp, and stubble
3. Animal waste	Dung, urine, and biogas slurry
Urban solid waste	Kitchen waste from household and restaurants, waste from market yards and places of worship, and sludge from Sewage treatment plants (STPs).
Agro-industries waste	
1. Food-processing unit	Peel, rind, and unused pulp of fruits and vegetables
2. Vegetable oil refineries	Press mud and seed husk
3. Sugar factories	Press mud, fine bagasse, and boiler ash
4. Breweries and distillery	Spent wash, barley waste, yeast sludge
5. Seed-production unit	Core of fruits, paper, and date-expired seeds
6. Aromatic oil extraction	Stems, leaves, and flowers after extraction of oil
7. Coir industries	Coir pith

Selection of Site: Any place with shade, high humidity and cool, a thatched roof to protect from direct sun light and rain. The waste can be covered with moist gunni bags.

Structure for Vermicompost: A cement tub with a height 2.5 feet and breadth of 3 feet, the bottom of which is to be made as a slope-like structure to drain the excess water from the unit, a small sump to collect the drain water. It can also be prepared in wooden boxes, plastic buckets, or in any containers with drain holes at the bottom (Fig. 2.7).

Feeding the Waste into the Structure: Waste material should be mixed with 30% cattle dung either by weight or volume. Mixed waste is placed into the tub/container up to the rim, the moisture level should be maintained at 60%. Over the waste material the selected earthworms are placed uniformly, for 1 meter cube space, 1 kg of worms (1000 number is required) (Fig. 2.8).

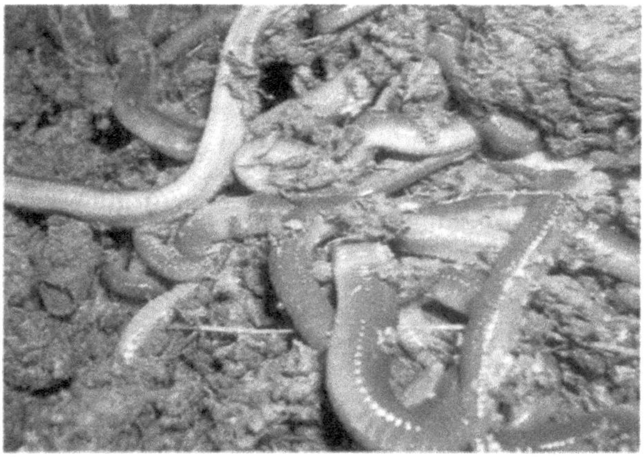

Figure 2.7 Earthworm suitable for vermicomposting.

Figure 2.8 Asian worms (*Perinonyx excavates*).

Figure 2.9 Earthworm suitable for vermicomposting.

Bedding Material: Sawdust or husks in about 3-cm layers, with fine sand and garden soil, is provided as bedding material inside the container. The waste material should be decomposed partially before introducing into the unit, and this composting material are added above this bedding material. The worms feeding actively assimilate only 5–10% for their growth and the rest is excreted as loose granular mounds or worm cast. A bed of 1 m × 1 m × 0.3 m requires 30–40 kg of bedding and feeding materials. This can support 1000–1500 earthworms, which would multiply and compost the matter from the upper layers. The first lot of vermicompost is ready in only 30–40 days. As per estimates available, 1 kg of earthworms (1000 adult numbers) would produce 10 kg casts in 60–70 days.

Watering of Vermibed: Daily pouring of water is not required for the vermibed, but 60% moisture should be maintained throughout the period. Watering should be stopped before the harvest of vermicompost (Fig. 2.9).

Precautionary Measures: Moisture level should be maintained around 50–60%. Temperature should be maintained within the range of 20–30°C. Handle the earthworms gently to avoid injury and protect from predators like ants, rats, etc. (Fig. 2.10).

Enriching of Vermicompost: It can be enriched with beneficial microorganisms like *Azotobacter*, *Azospirillum*, *Phosphobacteria*, and *Pseudomonas*. For 1 tonne of waste processing, 1 kg of azophos, which contains both *Azospirillum* and *Phosphobacteria*, should be inoculated 20 days after putting the waste into the vermin bed (Fig. 2.11).

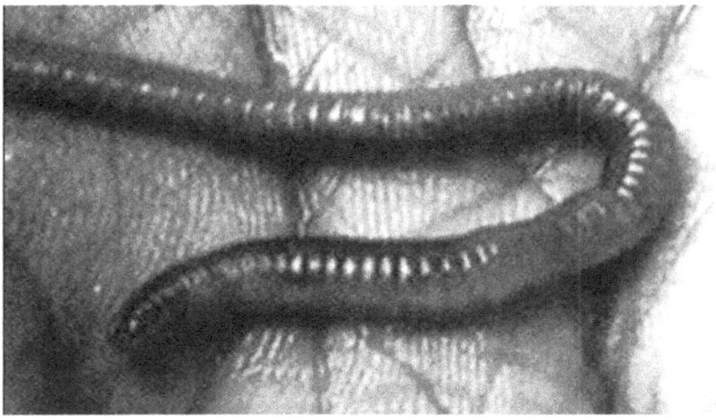

Figure 2.10 African earthworm (*Eudrillus euginiae*).

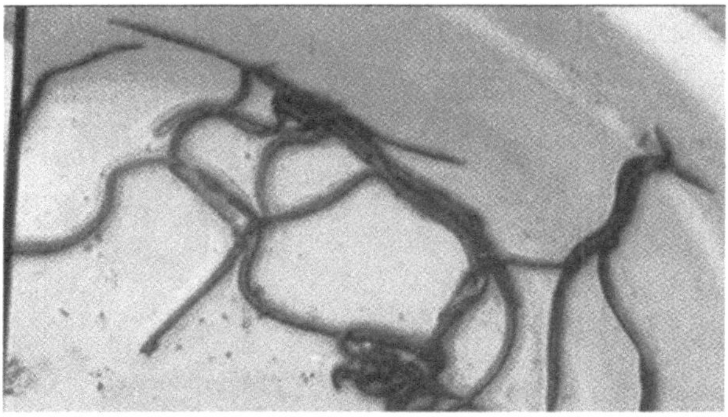

Figure 2.11 Tiger worm or red wrinkler (*Eisenia foetida*).

Harvesting of Vermicompost: In the tub method of composting, the casting formed on the top layer are collected periodically. The collection may be carried out once in a week. Periodically the casting will be scooped out and put in a shady place as a heaplike structure. The harvesting of castings should be limited up to the earthworm presence on the top layer.

Storing and Packing of Vermicompost: The harvested vermicompost (Fig. 2.12) should be stored in dark, cool place. It should have minimum 40% moisture, and sunlight should not fall over the composted material as it will (Fig. 2.13) lead to loss of moisture and nutrients. It is advocated that the harvested composted (Fig. 2.14) material is openly stored rather than packed in over gunnies. Packing can be done at the time of selling. If it is stored in an open place, a periodical sprinkling of water may be done to maintain the moisture level and also to maintain a beneficial microbial moisture population. If it becomes necessary to store the material, laminated gunnies are used for packing as this will minimize the

Figure 2.12 Introduction of earthworm into the vermibed.

Figure 2.13 Pelleted vermicas.

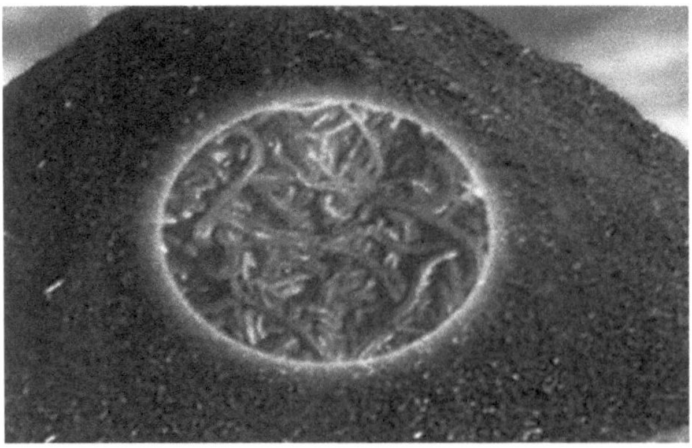

Figure 2.14 Harvested vermicompost.

moisture evaporation loss. Vermicompost can be stored for 1 year without loss of quality, if the moisture is maintained at 40% level.

Phases in vermicomposting

Phase 1 Processing involving collection of wastes. Shredding, mechanical separation of the metal, glass, and, ceramics, and storage of organic wastes.

Phase 2 Predigest ion of organic waste materials for 20 days by heaping the material along with cattle dung slurry. This process partially digests the material and makes it fit for earthworm consumption. Cattle dung and biogas slurry may be used after drying. Wet dung should not be used for vermicompost production.

Phase 3 Preparation of earthworm bed. A concrete base is required to put the waste for vermicompost preparation. Loose soil will allow the worms to go into soil and also while watering, all the dissolvable nutrients go into the soil along with water.

Phase 4 Collection of earthworms after vermicompost collection. Serving the composted material to separate fully composted material. The partially composted material will be again put into vermicompost bed.

Phase 5 Storing the vermicompost in a proper place to maintain moisture and allow the beneficial microorganisms to grow.

Nutritive Value of Vermicompost: The nutrient content in vermicompost varies depending on the waste materials that are being used for compost preparation. If the waste materials are heterogeneous, there will be a wide range of nutrients available in the compost. If the waste materials are homogenous, only certain nutrients are available. The common available nutrients in vermicompost are as follows:

Organic carbon (%) = 9.50–17.9
Nitrogen (%) = 0.50–1.50
Phosphorus (%) = 0.10–0.30
Potassium (%) = 0.15–0.56
Sodium (%) = 0.06–0.30
Calcium and magnesium
(mg/kg/100 g) = 22.7–47.6
Copper (mg/kg) = 2.00–9.50
Iron (mg/kg) = 2.00–9.30
Zinc (mg/kg) = 5.70–11.5
Sulfur (mg/kg) = 128–548

Benefits of Vermicompost: Advantages of vermicompost are it is a natural fertilizer prepared from biodegradable organic wastes and it is free from chemical inputs. There are no adverse effects on soil, plants, or the environment. Neutralization of highly acidic and alkaline soil is achieved.

It improves soil aeration and texture thereby reducing soil compaction. It improves water retention capacity of soil because of its high organic matter content, promotes better root growth and nutrient absorption, improves nutrient status of soil, both macronutrients and micronutrients. Worms are also used as pet food or fishing bait and hence have a commercial value. Vemicomposts have higher nutritive value when compared to biocompost. It has higher number of beneficial organisms like *Azotobacter*, *Azosprillum*, and *phosphobacteria*, these beneficial organisms contribute their benefits to the vermicompost.

2.7.3 Incineration

This is the process of direct burning of wastes in the presence of excess air (oxygen) at temperatures of about 800°C and above, liberating heat energy, inert gases, and ash. Net energy yield depends upon the density and composition of the waste; relative percentage of moisture and inert materials, which add to the heat less; ignition temperature, size, and shape of the constituents; design of the combustion system, i.e., fixed bed/ fluidized bed), etc. In practice, about 65−80% of the energy content of the organic matter can be recovered as heat energy, which can be utilized either for direct thermal applications, or for producing power via steam-turbine generators (with typical conversion efficiency of about 30%).

The combustion temperatures of conventional incinerators fueled only by wastes are about 760°C in the furnace, and in excess of 870°C in the secondary combustion chamber. These temperatures are needed to avoid odor from incomplete combustion but are insufficient to burn or even melt glass. To avoid the deficiencies of conventional incinerators, some modern incinerators utilize higher temperatures of up to 1650°C using supplementary fuel. These reduce waste volume by 97% and convert metal and glass to ash.

Waste burned solely for volume reduction may not need any auxiliary fuel except for startup. When the objective is steam production, supplementary fuel may have to be used with the pulverized refuse, because of the variable energy content of the waste or in the event that the quantity of waste available is sufficient.

While incineration is extensively used as an important method of waste disposal, it is associated with some polluting discharges that are of environmental concern, although in varying degrees of severity. These can fortunately be effectively controlled by installing suitable pollution control devices and by suitable furnace construction and control of the combustion process.

Basic Types of Incineration Plants: Both fixed-bed and fluidized-bed type furnaces are used in incinerators. The modern municipal incinerators are usually of the continuously burning type, and may have water-wall units that can, however, be a problem. Recent advancement include the twin-interchanging fluidized-bed combustor developed by a company in Japan, which is claimed to be capable of completely combusting wastes of low or high calorific values at very high overall efficiency. Some basic types of incineration plants operating in the developed countries in the West and in Japan are as follows:

2.7.3.1 Mass Burn

About three-fourths of the waste-to-energy facilities in the United States and a few other countries are "mass burn," where refuse is burned just as it is delivered to the plant. Without processing or separation. These plants are sized to incinerate up to 3000 tons of refuse per day and use two or more burners in a single plant. While facilities are sized according to the expected volume of waste, they are actually limited by the amount of heat produced when the garbage is burned. Some mass burn plants remove metals from the ash for recycling. Mass burn plants have operated successfully in Europe for more than 100 years.

2.7.3.2 Modular Combustion Unit

Modular incineration are simply mass-burn plants with capacity ranging from 25 to 300 tonnes/day. The boilers are built in a factory and shipped to the plant site rather than being erected on the site, as is the case with larger plants. These facilities are often used in small communities.

2.7.3.3 Refuse-Derived Fuel (RDF) Based Power Plants

In an RDF plant, waste is processed before burning. Typically the non-combustible items are removed, separating glass and metals for recycling. The combustible waste is shredded into a smaller, more uniform particle size for burning. The RDF thus produced may be burned in boilers onsite or it may be dipped to offsite; if it is to be used offsite it is usually densified into pellets through the process of pelletization.

Pelletization involves segregation of the incoming waste into high and low calorific value materials and shredding them separately, to nearly uniform size. The different heaps of the shredded waste are then mixed together in suitable proportion and then solidified to produce RDF Pellets. The calorific value of RDF pellets can be around 4000 kcal/kg depending upon the percentage of organic matter in the waste, additives, and binder materials used in the process, if any. Since pelletization enriches the organic content of the waste through removal of inorganic materials and moisture, it can be a very effective method for preparing an enriched-fuel feed for other thermochemical processes like pyrolysis/gasification, apart from incineration. An additional advantage is that the pellets can be conveniently stored and transported.

RDF plants involve significantly more sorting and handling than mass-burn facilities and therefore provide greater opportunity to remove environmentally harmful materials from the incoming waste prior to combustion. However, it is not possible to remove the harmful materials completely. Several years ago RDF was used mainly along with coal-fired boilers but now, because of the stricter restrictions with reference to air emissions, it is usually burned in dedicated boilers designed and built specially for the RDF. In case of RDF pellets too, it needs to be ensured that the pellets are not burned indiscriminately or in the open, but only in dedicated boilers.

All sorts of waste materials are generated in the Indian cities as in other countries. However, in the absence of a well-planned, scientific system of waste management (including waste segregation at source) and of any effective regulation and control of rag-picking, waste-burning, and waste-recycling activity, the leftover waste at the dumping yards generally contains a high percentage of inerts (>40%) and of putrescible organic matter (30–60%). It is common practice of adding the road sweepings to the dustbins. Papers and plastics are mostly picked up and only such fraction that is in an unrecoverable form remains in the refuse. Paper normally constitutes 3–7% of refuse while the plastic content is normally less than 1%. The calorific value on dry weight basis (high calorific value) varies between 800 and 1100 kcal/kg. Self-sustaining combustion cannot be obtained for such waste and auxiliary fuel will be required.

However, with the growing problems of waste management in the urban areas and the increasing awareness about the ill effects of the existing waste management practices on the public health, the urgent need for improving the overall waste management system and adoption of advanced, scientific methods of waste disposal, including incineration, is imperative.

2.7.3.4 Pyrolysis/Gasification

Pyrolysis is also referred to as destructive distillation or carbonization. It is the process of thermal decomposition of organic matter at high temperature, about 900°C.

In an inert (oxygen-deficient) atmosphere or vacuum, it is producing a mixture of combustible carbon monoxide, methane, hydrogen, ethane [CO, CH_4H_2O H_2 C_2 H_6], and noncombustible carbon dioxide, water, nitrogen gases, pyroligenous liquid, chemicals, and charcoal. The pyroligenous liquid has high heat value and is a feasible substitute of industrial fuel oil. The amount of each end-product depends on the chemical composition of each product and changes with pyrolysis temperature, residence time, pressure, feedstock, and other variables.

Gasification involves thermal decomposition of organic matter at high temperatures in the presence of limited amounts of air/oxygen, producing mainly a mixture of combustible gas (carbon monoxide, hydrogen, and carbon dioxide). This process is similar to pyrolysis, involving some secondary/different high-temperature (>1000°C) chemistry, which improves the heating value of gaseous output and increases the gaseous yield (mainly combustible gases $CO + H_2$) and lesser quantity of other residues. The gas can be cooled, cleaned, and then utilized in internal combustion (IC) engines to generate electricity.

Pyrolysis/gasification is already a proven method for homogenous organic matter like wood, pulp, etc., and is now being recognized as an attractive option for municipal solid waste also. In these processes, besides net energy recovery, proper destruction of the waste is also ensured. The products are easy to store and handle. These processes are therefore being increasingly favored in place of incineration.

Different Types of Pyrolysis/Gasification System: The salient features of different types of pyrolysis/gasification systems so far developed are given below.

Garret's Flash Pyrolysis Process: This low-temperature pyrolysis has been developed by Garret Research and Development Company. In a 4 tonnes/day pilot plant set up by the company at La Verne, California, the solid waste is initially coarsely shredded to less than 50 mm size, air classified to separate organics/inerts, and dried through an air drier. The organic portion is then screened, passed through a hammer mill to reduce the particle size to less than 3 mm, and then pyrolyzed in a reactor at atmospheric pressure. The proprietary heat-exchange system enables pyrolytic conservation of the solid waste to a viscous oil at 500°C.

Pyrolysis Process Developed by Energy Research Center of Bureau of Mines, Pittsburgh: This is a high-temperature pyrolysis process to produce both fuel oil and fuel gas and has been investigated mainly at laboratory scale. The waste charge is heated in a furnace with nickel–chromium resistors to the desired temperature. The produced gases are cooled in an air trap where tar and heavy oil condense out. Uncondensed vapors pass through a series of water-cooled condensers where additional oil and aqueous liquors are condensed. The gases are then scrubbed in an electrostatic precipitator before further use. It is claimed that 1 tonne of dried solid waste produces 300–500 m^3 of gas, but the process is yet to be tested at full scale.

Destrugas Gasification System: In this system the raw solid waste is first subjected to shredding/size reduction in an enclosed shed. The air from this shed is taken up as intake air in the plant so as to avoid odor problems. The shredded waste is fed to retorts (heated indirectly by burning gas in a chamber enveloping it) through which it sinks under gravity and gets subjected to thermal decomposition. The produced gas is washed and most of it (85%) used for heating the retorts. The remaining 15% is available as fuel. The slag consists of mostly char.

2.7.3.5 Slurry Carb Process

This process has been developed by a company in the United States to convert municipal solid waste into fuel oil. It is used in conjunction with a wet resource-recovery process to separate out the recyclables. The received waste is first shredded and placed in an industrial pulper. The heavier and denser inorganic material sinks to the bottom of the water-filled pulper, from which it is easily removed. The remaining waste slurry (organic fraction) is subjected to violent pulping action, which further reduces the size of its constituents. The pulped organic waste is then subjected to high pressure and temperature whereby it undergoes thermal decomposition/carbonization (slow pyrolysis) to fuel oil.

2.7.3.6 Plasma Pyrolysis Vitrification (PPV)/Plasma Arc Process

This is an emerging technology utilizing thermal decomposition of organic wastes for energy/resource recovery. The system basically uses a plasma reactor, which houses one or more plasma arc torches that generate, by application of high voltage between two electrodes, a high-voltage discharge and consequently has an extremely high temperature environment (between 5000 and 14,000°C). This hot plasma zone dissociates the molecules in any organic material into the individual elemental atoms while all the materials are simultaneously melted into molten lava.

The waste material is directly loaded into a vacuum in a holding tank, preheated, and fitted to a furnace where the volatile matter is gasified and fed directly into the plasma-arc generator where it is preheated electrically and then passed through the plasma arc dissociating it into elemental stages. The gas output after scrubbing comprises mainly of CO and H_2. The liquefied produce is mainly methanol.

The entire process is claimed to safely treat any type of hazardous or nonhazardous materials. It has the advantage that the NO_x (oxides of nitrogen) and SO_x (oxides of sulfur) gases emissions do not occur as in normal operation due to the lack of oxygen in the system.

Advantages and Disadvantages of Different Technological Options: The main advantages and disadvantages of the different technological options described are given in the following Table.

Advantages	Disadvantages
Anaerobic digestion	
Energy recovery with production of high-grade soil conditioner.	Heat released is less, resulting in lower and less-effective destruction of pathogenic organisms than in aerobic composting. However, now thermophilic temperature systems are also available to take care of this.
No power requirement unlike aerobic composting, where sieving and turning of waste pile for supply of oxygen is necessary.	
Enclosed system enables all the gas produced to be collected for use. Controls greenhouse gas emissions.	Unsuitable for wastes containing less organic matter.
Free from bad odor, rodent, and fly menace, visible pollution and social resistance.	Requires waste segregation for improving digestion efficiency.
Modular construction of plant and closed treatment needs less land area.	
Net positive environmental gains.	
Can be done at small scale.	
Landfill gas recovery	
Least cost option.	Generally polluted surface runoff during rainfall.
The gas produced can be utilized for power generation or as domestic fuel for direct thermal applications.	
Highly skilled personnel not necessary.	Soil/ground water acquirers may get contaminated by polluted leachate in the absence of proper leachate-treatment system.
Natural resources are returned to soil and recycled.	

(Continued)

Advantages	Disadvantages
Can convert low-lying marshy land to useful areas.	Inefficient gas recovery process yielding 30–40% of the total gas generation. Balance gas escapes to the atmosphere (significant source of two major greenhouse gases, namely carbon dioxide and methane).
	Large land-area requirement.
	Significant transportation costs to faraway landfill sites may upset viability.
	Cost of pretreatment to upgrade the gas-to-pipeline quality and leachate treatment may be significant.
	Spontaneous ignition/explosions due to possible buildup of methane concentrations in atmosphere.
Incineration	
Most suitable for high-calorific-value waste, pathological wastes, etc.	Least suitable for aqueous/high-moisture-content/low-calorific-value, and chlorinated waste.
Units with continuous feed and high throughput can be set up.	Excessive moisture and inert content affects net energy recovery; auxiliary fuel support may be required to sustain combustion.
Thermal energy recovery for direct heating or power generation.	
Relatively noiseless and odorless.	Concern for toxic metals that may concentrate in ash, emission particulates, SO_x, NO_x, chlorinated compounds, ranging from HCI to dioxins.
Low land area requirement.	
Can be located within city limits, reducing the cost of waste transportation.	High capital and Operation and maintenance (O & M)costs.
	Skilled personnel required for O & M.
Hygienic.	Overall efficiency low for small power stations.
Pyrolysis/gasification	
Production of fuel gas/oil, which can be used for a variety of applications	Net energy recovery may suffer in case of wastes with excessive moisture.
Compared to incineration, control of atmospheric pollution can be dealt with in a superior way, in a technoeconomic sense.	High viscosity of pyrolysis oil may be problematic for its transportation and burning.

Land Requirements: The area of land required for setting up any waste-processing/treatment facility generally depends upon the following factors:

- total waste-processing and waste-treatment capacity, which will govern the overall plant design/size of various subsystems;
- waste quality/characteristics, which will determine the need for pre-processing, if required, to match with the plant design;
- waste-treatment technology selected, which will determine the waste fraction destroyed/converted to energy;
- quantity and quality of reject waste, liquid effluents, and air emissions, which will determine the need for disposal/post treatment requirements to meet Environmental pollution control (EPC) norms.

As such, the actual land-area requirement can be worked out only in the detailed project report for each specific project. However, for initial planning the following figures may be considered for 300 tonnes per day (TPD) (input capacity) waste-to-energy facilities:

Incineration/gasification/pyrolysis plants:	0.8 hectare
Anaerobic digestion plants:	2 hectares
Sanitary landfills (including gas-to-energy recovery):	36 hectares

Utilization of Biogas: Main constituents of biogas are methane (about 60%), carbon dioxide (about 40%) and small quantities of ammonia and hydrogen sulfide. The calorific value of biogas is about 5000 kcal/m^3 and depends upon the methane percentage. The gas from landfills generally has a lower calorific value. The biogas, by virtue of its high calorific value, has tremendous potential to be used as fuel for power generation through either IC engines or gas turbines.

Local Gas Use: The simplest and most cost-effective option for use of landfill gas/biogas is local gas use. This option requires that the gas be transported, typically by dedicated pipeline, from the point of collection to the points(s) of gas use. If possible, a single point of use is preferred so that pipeline construction and operation costs can be minimized. Prior to transporting the gas to the user, the gas must be cleaned to some extent; condensate and particulates are removed through a series of filters and/or driers. Following this minimal level of gas cleaning, gas quality of 35–50% methane is typically produced. This level of methane concentration is generally acceptable for use in a wide variety of equipment, including boilers and engines. Although the gas-use equipment is usually designed to handle natural gas that is nearly 100% methane, the equipment can usually be adjusted easily to handle the gas with the lower methane content.

Pipeline Injection: Pipeline injection may be a suitable option if no local gas user is available. If a pipeline carrying medium quality gas is nearby, only minimal gas processing may be needed to prepare the gas for injection. Pipeline injection requires that the gas be compressed to the pipeline pressure.

- *Medium-Quality Gas*: Medium-quality gas will typically have an energy value that is the equivalent to landfill gas with a 50% methane concentration. Prior to injection, the gas must be processed so that it is dry and free of corrosive impurities. The extent of gas compression and the distance required to reach the pipeline are the main factors affecting the attractiveness of this option.
- *High-Quality Gas*: For high-quality gas, most of the carbon dioxide and trace impurities must be removed from the recovered gas. This is a more difficult and hence more expensive process than removing other contaminants.

Electricity Generation: Electricity be generated onsite or for distribution through the local electric power grid.

Internal Combustion (IC) Engines: IC engines are the most commonly used conversion technology in landfills gas applications. They are stationary engines, similar to conventional automobile engines, that can use medium-quality gas to generate electricity. While they can range from 30 to 2000 kilowatts (KW), IC engines associated with landfills typically have capacities of several hundred KW.

IC engines are a proven and cost-effective technology. Their flexibility, especially for small generating capacities, makes them the only electricity-generating option for smaller landfills. At the start of a recovery project, a number of IC engines may be employed; they may then be phased out or moved to alternative utilization sites, as gas production drops.

IC engines have proven to be reliable and effective generating devices. However, the use of landfill gas in IC engines can cause corrosion due to the impurities in landfill gas. Impurities may include chlorinated hydrocarbon that can react chemically under the extreme heat and pressure of an IC engine. In addition, IC engines are relatively inflexible with regard to the air-fuel ratio, which fluctuates with landfill gas quality. Some IC engines also product significant NO_x emissions, although designs exist to reduce NO_x emissions.

- *Gas Turbines*: Gas turbines can use medium-quality gas to generate power of sale to nearby users or electricity supply companies, or for onsite use. Gas turbines typically require higher gas flows than IC engines in order to be economically attractive, and have, therefore, been used at larger landfills. They are available in sizes from 500 KW to 10 MW, but are most useful for when they are 2–4 MW. Gas turbines consume approximately the same amount of fuel when generating power. Additionally, the gas must be compressed prior to use in the turbine.
- *Steam Turbines*: In cases where extremely large gas flows are available steam turbines can be used for power generation.
- *Fuel Cell*: Fuel cells, an emerging technology, are being tested with landfill gas. These units, expected to be produced in the 1–2 MW capacity range, are highly efficient with relatively low NO_x emissions. They operate by converting chemical energy into usable electric and heat energy.

Purification of Biogas: Most effluents and solid wastes contain sulfates, which give rise to the presence of H_2S content of up to 1000 ppm, beyond which the H_2S can cause rapid corrosion. Although biogas generated from municipal solid waste is generally not expected to contain a high percentage of H_2S, adequate arrangements for cleaning of the gas have to be made in case it is beyond 1000 ppm. Systems being used to remove H_2S from biogas are based on chemical, biochemical, or physical processes (Fig. 2.15).

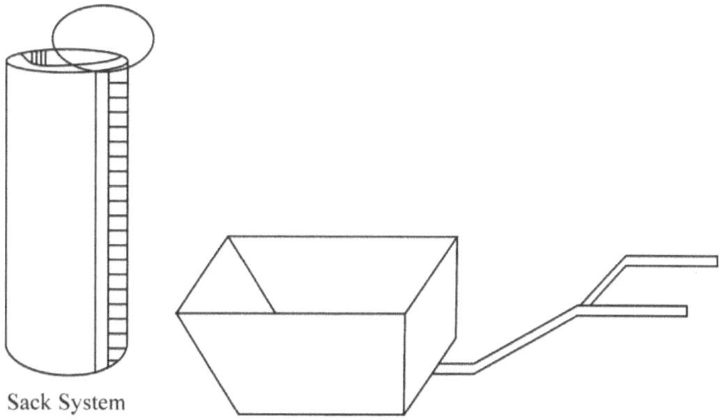

Figure 2.15 Two-wheeled wheelbarrow.

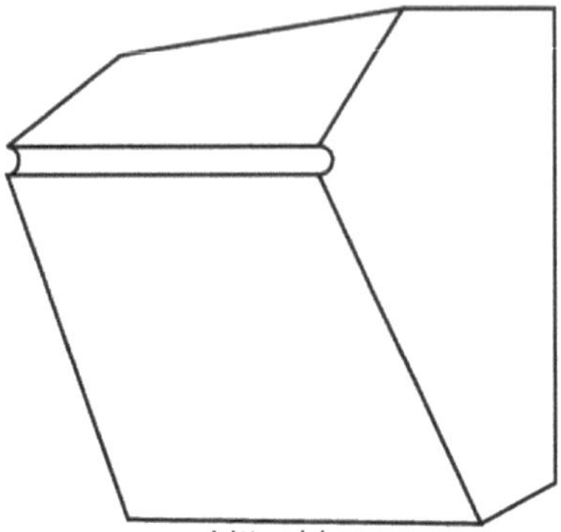

Figure 2.16 Litter bin.

Chemical Processes: Chemical processes are based on absorption of H_2S by alkali, iron or amines. The most widely used process for desulfurization is the amine process because it selectively absorbs H_2S from biogas and can be carried out at near atmospheric pressure. This can reduce the H_2S content to 800 ppm. The raw biogas is treated through an absorber column against triethanol amine solution. The absorber has one or more packing beds of polypropylene rings to provide better contact between gas and the liquid media. The amine solution, while reacting with biogas, gets saturated with H_2S and CO_2 and is sent to the stripper column wherein it gets regenerated by stripping off the H_2S and CO_2 by heating with steam. The sour gases are let off to a chimney. The regenerated amine is ready for reuse (Fig. 2.16).

Biochemical Processes: These processes use secondary treated effluent to clean the biogas. This effluent is sprayed from the top of the absorber columns while the raw biogas is blown in from the bottom. The effluent cleans the biogas and is then sent to the aeration tank where the H_2S is converted into sulfates. The effluent from the aeration tank is partly supplemented by fresh treated effluent and partly disposed of. The formation of elemental sulfur is outside the scrubber and therefore there is availability of the scrubber without choking effect (Fig. 2.17).

Summary of Gas-Cleaning Methods: A summary of the different methods being used for purification of biogas is given in the following Table 2.12.

2.8 TOOLS AND EQUIPMENT

For a successful solid waste management, use of appropriate tools and equipment is very essential. This section deals with pictorial representation of the various equipment used for the collection, storage, and final

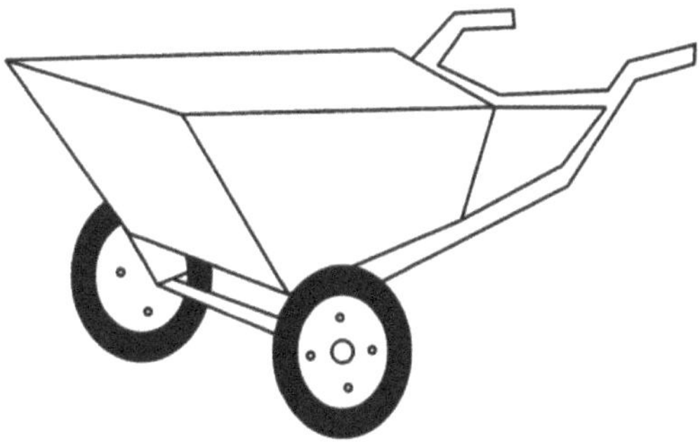

Figure 2.17 Three-wheeled wheelbarrow.

Table 2.12 Summary of gas cleaning methods

Compound	Process type	Process alternative available
H_2O	Absorption	1. Silica gel 2. Molecular sieves and alumina
	Adsorption	1. Ethylene glycol (at low temperature of −20°F) 2. Selexol
	Refrigeration	Chilling to 4°F
Hydrocarbons	Adsorption	Activated carbon
	Absorption	1. Lean oil absorption 2. Ethylene glycol 3. Selexol (All at low temperature of −20°F)
	Combination	Refrigeration with ethylene glycol plus activated carbon absorption
CO_2 and H_2S	Absorption	1. Organic solvents 2. Alkaline salt solutions 3. Alkanolamines
	Adsorption	1. Molecular sieves 2. Activated carbon
	Membrane separation	Hollow fiber membrane
	H_2S removal with sulfur recovery	Biochemical process

disposal of solid wastes. It further depicts the advancement made in the use of various equipment for a safe solid waste management. The pictures in this section cover almost all the fields of solid waste management like onsite storage, collection, transport, disposal, and resource recovery (Figs. 2.18 and 2.19).

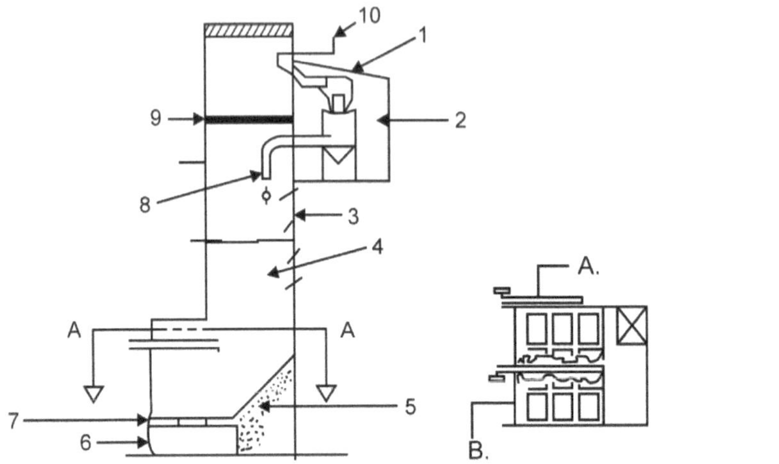

(A) Outside over fire air manitord and fan. (B) Inside over fire air manitord and fan.

1. Washer 2. Washer and induced draft fan 3. Hopper door locks system 4. Charging and gas flue 5. Steep hearth 6. under fire register 7. Enarged great area 8. Gas Inlet to washer 9. By pass damper with remote control.

Figure 2.18 (A) Outside overfire air manifold and fan. (B) Inside overfire air manifold and fan. (1) Washer. (2) Washer and induced-draft fan. (3) Hopper door locks system. (4) Charging and gas flue. (5) Steep hearth (6) Underfire register. (7) Enlarged great area. (8) Gas inlet to washer. (9) Bypass damper with remote control.

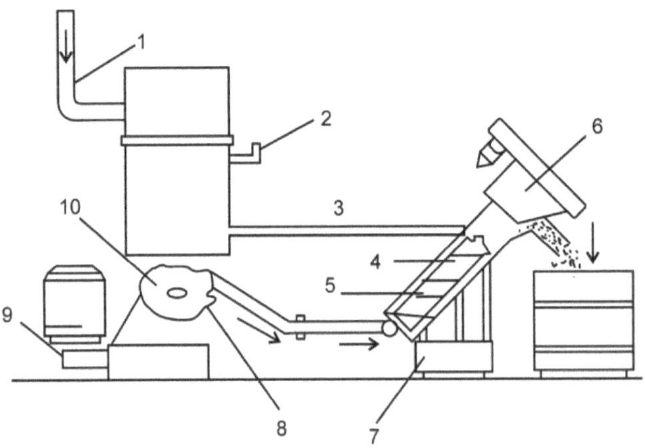

1. Solid Waste 2. Fresh Water Inlet 3. Return Water 4. Perforated Screen 5. Hellical Screw 6. Compression Ring 7. Slurry Return Pump 8. Sizing Ring 9. Slurry Pump

Figure 2.19 Single-flue incinerator. (1) Solid waste. (2) Freshwater inlet. (3) Return water. (4) Perforated screen. (5) Helical screw. (6) Compression ring. (7) Slurry return pump. (8) Sizing ring. (9) Slurry pump.

2.9 RECLAMATION, REUSE, AND RECOVERY OF ENERGY FROM MUNICIPAL SOLID WASTE

The term "recovery" implies the physical separation of a component such as ferrous metals from the mixed waste, while "reclamation" implies the chemical transformation of the waste into a new product such as a fuel.

The majority of the processes used for reclamation of municipal refuse are physical means, while the energy recovery is carried out through various processes like burning raw refuse in steam-generating incinerators, pyrolysis, hydrogenation, controlled anaerobic digestion, and recovery of methane from landfills.

The physical separation process of municipal refuse for reclamation is further divided into:

- primary separation of waste components,
  e.g., separating organic and inorganic fraction
- secondary separation of particular components
  e.g., magnetic separation of ferrous metals from the inorganic fraction
- tertiary separation, used to upgrade separated fractions
  e.g., separation of glass from contaminants such as ceramics, stone, and bone

The organic fraction is further processed for energy recovery, while the inorganic fraction can be used for landfill or may be processed further for materials recovery.

2.9.1 Separation of Organic and Inorganic Fractions

There are three methods for the primary separation of municipal refuse:

- wet pulverization
- wet pulping
- dry separation, involving size reduction in a hammer mill

2.9.2 Wet Pulverization

The process involves the untreated waste being moisturized and fed into a large, slowly rotating drum, in which self-pulverization is achieved by tumbling action of hard components. The waste, which is separated into fine and coarse fractions, is further processed by passing a fine fraction through screens and the resultant fraction is mainly organic while the coarse fraction passing out is mainly inorganic.

2.9.3 Wet Pulping

Here the waste is introduced as an aqueous slurry (3–10% solids) and is reduced in size by a fast-rotating segmented blade. The pulpable waste comes out of the bottom of the pulper. This pulped waste is largely organic in nature while the inorganic waste removed in a liquid cyclone by centrifugal action.

2.9.4 Dry Separation

This method is the most commonly used method for solid waste recovery or reclamation.

The major process involved are

- size reduction
- screening
- air classification
- magnetic separation

Further the above processes are modified to get a rich waste consisting of paper and plastic, which can be used as fuel or for reclamation purposes.

2.9.5 Secondary Separation

This process used for separating particular components is carried out by the following one or more processes:

- magnetic separation
- screening, with the oversize containing metals and fines most of the glass
- a mineral jig that separates waste particles by specific gravity
- a rising current or hydraulic classification
- a heavy-media separator, in which the choice of density and particle size of the feedstock results in efficient separation of organics, from glass and metals, of glass from metals, of different metals, or of ceramics from glass

2.9.6 Tertiary Separation

This method for inorganic waste is aimed at recovery of nonferrous metal or glass fractions for recycling. The three basic alternatives are:

- heavy media separation
- electrodynamic or eddy current separation specially for aluminum
- electrostatic separation

The energy recovery is carried out by the following processes:

1. burning of raw refuse for steam generation
2. burning prepared refuse (RDF) in modified, existing or new steam generators
3. pyrolysis, hydrogenation
4. controlled anaerobic digestion
5. recovery of methane from landfills

Burning of Raw Refuse for Steam Generation: In this process the collected waste heaps are mixed by cranes to get a homogenous mixture and then fed to furnaces. The most common furnace used is of the water-wall type. The walls of the combustion chamber consist of hundreds of feet of tubing with water circulating through them. The heat produced converts water to steam. The furnace should be operated at about 1000°C by which odds can be eliminated from burning refuse and prevent the ash from fusing and fouling the water tubes. Since the municipal refuse is lower in sulfur content, the problem of sulfur oxide emissions are avoided. However, the formation of hydrochloric acid due to chlorine-bearing material in the refuse should be minimized by using wet scrubbers.

Two major methods of producing oil from organic material are:

- pyrolysis
- hydrogenation

Pyrolysis: This is defined as the thermal degradation of organic material in an oxygen-deficient atmosphere, which results in liquid fuel. Here the refuse is separated into an organic fraction after shredding, with a moisture content of 3%. Further, after removing ferrous material magnetically, pyrolysis takes place in a reactor where the organic fraction is mixed with burning char. The temperature is maintained at 480°C, which prevents gaseous products from thermal degradation. After passing hot gases through cyclones to remove char, oil is formed by rapidly cooling the gases in a venture-quench system. The main disadvantage of this process to pyrolysis is that it needs pressure reactors.

Gaseous Fuel: Gas can be recovered by pyrolysis, gasification, or anaerobic digestion.

Recovery of Methane from Sanitary Landfills: This process was first practiced in Los Angeles in the United States. It was found that methane gas produced by natural anaerobic decomposition of waste in a landfill is 50–55% of the total gas coming out in a landfill. The landfill gas is passed through a series of towers containing clay material absorbent medium to remove moisture and CO_2. The remaining gas is of pure methane and is pressurized and piped into an existing natural gas pipeline system.

Electricity: Electricity can be generated from municipal solid waste by either producing steam and using turbines or by using exhaust combustion gases directly in a gas turbine.

Resource-Recovery Strategies: In practice, the current strategy is to either use landfill methods or incinerate the refuse. Both approaches are disposal-volume strategies only, with the incineration providing reduction at the expense of air pollution and higher capital and operating costs. Landfill is truly a land-reclamation process. The alternative strategies involving resource recovery can be categorized in four general areas:

- material recovery
- energy recovery
- land recovery
- combinations of the above

Material Recovery: Material recovery can be performed by dry separation system, in which the valuable products are recovered by a system of dry shredding and air, magnetic, and electrostatic separation techniques. Incinerator residue is processed for recovery of metals using mineral beneficiation techniques common to the mining industry.

Energy Recovery: Energy recovery of refuse can be accomplished by a variety of techniques. The flow sheets (Fig. 2.20) for the recovery of resources, conversion products, and energy from solid wastes are shown in following figures. The most straight forward in concept is to use a water-wall incinerator as a boiler for steam recovery. This technique requires a nearby user of steam. In Europe, the steam generated is used in onsite turbines to generate electric power.

Another energy-recovery approach is to convert the refuse into fuel to be consumed elsewhere. One approach is to shred or wet-pulp the refuse and then burn it as a supplemented fuel in an electrical utility boiler. When used as an auxiliary fuel, the corrosive effects and particulate emissions from refuse burning are diluted by the primary fuel to more acceptable levels. This is possible when burning refuse alone.

An alternative to preparing fuel by shredding or wet pulping can be found in various pyrolysis processes. Pyrolysis can generally be classified as the thermal breakdown of material in the absence or near-absence of oxygen, such as can be found in coke and charcoal ovens. Recent developments of pyrolysis of solid waste include processes to produce an oil and to produce various forms of low-BTU fuel gases.

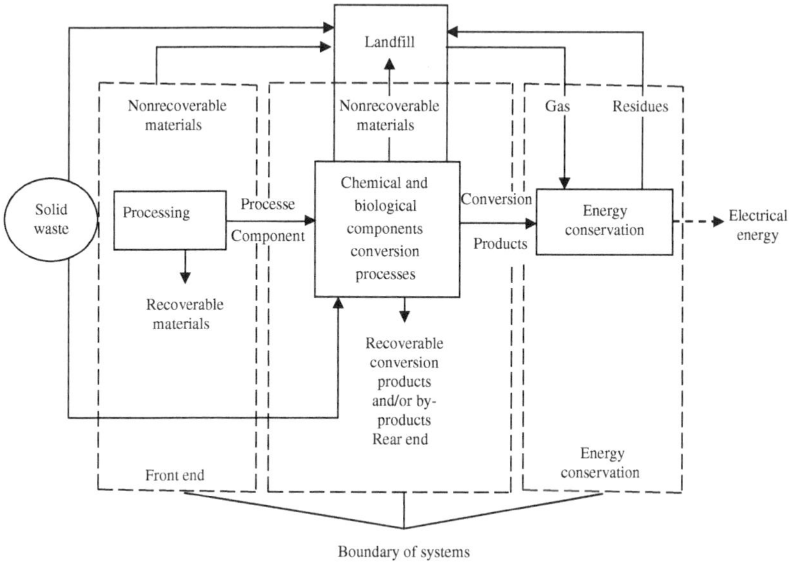

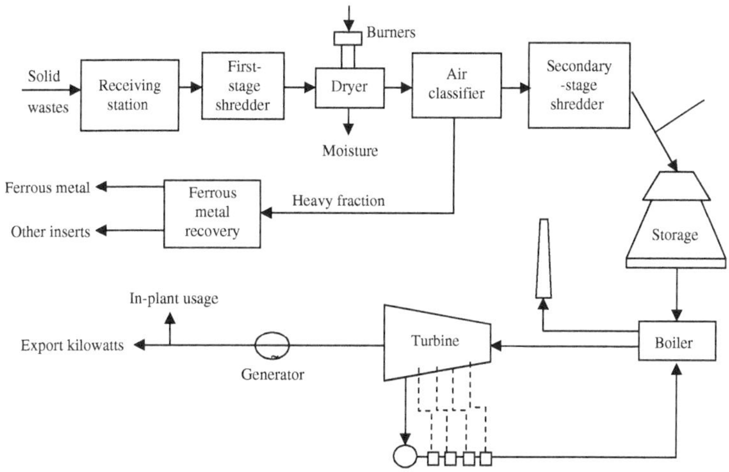

Figure 2.20 Typical flow sheet for recovery of resources, conversion products, and energy from solid wastes. *Source: Lunn, Low, Tom and Hara Inc. and Metcalf & Eddy Engineers, Inc.*

Sanitary landfilling is regarded as one of the best methods available for solid waste disposal. It is generally neat, safe, and inexpensive. Decomposition of landfills depends on permeability of cover material, depth of burial, rainfall, moisture content, putrescibility of the waste, and the degree of compaction. Decomposition of solid waste produces various gases.

These gases include methane, carbon dioxide, nitrogen, oxygen, and hydrogen sulfide.

Indian municipal refuse normally has an in situ density of 1000 kg/cu.m and its organic fraction (which is 50% of refuse) contains 50–55% carbohydrates, 4.5–5% proteins and 0.1–1% fats.

Theoretically, on anaerobic decomposition, the amount of CO_2 and CH_4 evolved from one kg each of carbohydrates, proteins, and fats are 0.456, 0.516, 0.448 cu.m of CO_2 and 0.455, 0.548, 1.095 cu.m of CH_4 respectively.

Hence, after decomposition the yields of methane and CO_2 per kg of Indian city refuse are:

	CH_4	CO_2
Carbohydrates	$0.25 \times 0.455 = 0.1137$	$0.25 \times 0.456 = 0.114$
Proteins	$0.035 \times 0.548 = 0.0149$	$0.025 \times 0.516 = 0.0129$
Fats	$0.005 \times 1.095 = 0.0054$	$0.005 \times 0.448 = 0.0022$
Total	0.134 cu.m	0.129 m

Total amount of gas produced from refuse is $(0.134 + 0.129 = 0.263)$ cu.m/kg or 263 cu.m/t, assuming a solid waste production of 0.5 kg/capita/day for a city population of 300,000.

The amount of refuse produced in 20 years for the above city will be:

$$= 0.5 \times 3 \times 10^5 \times 365 \times 20 = 10.95 \times 10^5 \text{t}$$

The quantity of gas produced in anaerobic decomposition will be $= 10.95 \times 263 \times 10^3 = 288$ cu.m (containing 140 cu.m CH4).

Gas production rate during the 20 years will be:

$$= 288 \times \frac{10^6}{20 \times 365 \times 24 \times 60} = 27.4 \text{ cu.m/min}$$

The gas from the landfill will have to be removed with the help of a number of pumping wells. It is normally assumed that all the gas occurring within the radius of influence of the well will be drawn by the concerned well.

If landfill gas is to be tapped, it is necessary to control migration of gas by

- placing impermeable lining material at or beyond the boundary to prevent escape of gas
- selective placement of granular material at or beyond the land to the boundary for venting and/or collecting the gas
- evacuation and venting of gas from landfill itself
- evacuation and venting of gas from the perimeter area beyond landfill

The gas collected is normally subjected to a certain amount of pretreatment for the removal of water vapor, CO_2, and the impurities. These are normally absorbed on alumina gel and activated carbon. Then, the gas is compressed depending upon the design of the gas-transportation system and distance.

2.9.7 Recycling of Solid Wastes

Recycling and recovery of solids and other components from solid wastes is essential not only for pollution abatement but also for economical management of an industry. The choice between recovery of valuable materials and disposal of solid wastes depends on the following three factors: technology, economics, and attitude. It is difficult to generalize the procedure for recycling of solid wastes since the technology and economics depend on their characteristics, quantity, and site location. Though technology may not be a major problem as there are adequate physical and chemical methods available, economics is the most important factor and also the most difficult to analyze.

Most of the household, municipal, and industrial solid wastes are being disposed of indiscriminately and with utter disregard to various problems arising out of such actions. Environmental damage, economic considerations, and resource conservation are totally disregarded to various problems arising out of such actions. Environmental damage, economic considerations, and resource conservation are totally disregarded. In some cases, the viability of a recovery process is the high cost of recovering a low-value waste material and consequent unprofitability frequently determine the decision for recycling. However, with the growing public awareness of problems of environment and Pollution control board (PCB's) interest in implementation of the laws and regulations, as well as the rapidly increasing cost of landfill and transportations, the recycling of wastes is gaining preference over disposal methods. The third factor, attitude, is difficult to quantify. Even if the recovery and recycle of a particular solid waste is technically, economically, and commercially viable, it is ignored for various other reasons such as prejudice, politics, vested interests, etc. This is the most difficult area for positive comment, and the process of educating the concerned persons regarding the economic and environmental benefits is rather slow.

2.9.8 Recycling Versus Disposal

It is essential to determine the characteristics of the solid waste before choosing between two alternatives, i.e., recycling and disposal. Sometimes a waste may be subdivided, one part being recycled and the other disposed. Also, in an industry it is important to distinguish between onsite and offsite operations. In the former's case, the disposal contractor decides and assumes responsibility regarding the economics and ecological aspects of disposal or recycling. The wastes may be directly or indirectly recycled and also internally or externally. Examples of direct recycling are process scrap in industries such as iron and steel, glass, paper, plastics, textiles, etc. Similarly in the refractory industry a portion of the ground powder of broken fine brick is added to the virgin clay to improve the quality of the finished bricks. Indirect internal recycling includes reprocessing of specification materials again. Recovery of plating chemicals is one such example. External recycling also follows a similar pattern except that the producer may not have the facilities available to reprocess the material or find it more economical to process it outside. In all these recycling procedures the two main considerations in assessing the alternatives available are technical feasibility and economic viability, though in most cases the second factor is the deciding one.

There are also other factors that affect final decisions depending on individual characteristics. The factors are:

1. *Capital Availability*: Although indirect internal recycling may be an attractive technical and economic proposition, ready capital for this process may be difficult to organize.
2. *Economic Capacity*: It may not be economically viable to resort to internal recycling. It is therefore advisable to investigate the possibility of collaboration with other organizations for a similar activity.
3. *Research and Development*: To investigate the most economic proposition to handle waste, this factor may be considered if the quantum of waste is large.
4. *Cost–Benefit Analysis*: It may be necessary to prepare a cost–benefit analysis and even to assign artificial costs to certain wastes when considering social problems.
5. *Supply and Demand Positions*: When the raw materials are freely available the recovery and recycle of essential components from a waste may not be economically attractive, but when the raw materials get depleted and their costs keep spiraling up, then recycling becomes a must in every stage of operation.

2.9.9 Recovery and Recycle

1. Household and municipal solid wastes

Recovery and recycle of various materials from solid wastes may be classified as:

a. recovery of metals
b. recovery of less-common metals
c. recovery of nonmetals such as plastics, glass, paper, rubber, fibers, etc.
d. recovery of materials from household solid refuse
e. energy and biogas recovery from solid wastes

In Indian cities the percentage of metal content in solid wastes from household and municipal refuse is very little but industrial solid waste is greater. The major content of municipal and household solid wastes contain paper (10%) and rags (5%) and the metal content is often less than 1%. Even this metal content is only scrap or aluminum and not copper and of this most of the reusable articles such as containers, etc., get picked by urchins. Hence, for Indian conditions, the recovery cycle of metals is used only for industrial solid wastes. As regarding the recovery of less common metals from solid waste, except in major cities and towns near industrial locations of these materials, their content is very low. But if some of these materials are hazardous or toxic in nature, they should be recovered and properly disposed after suitable treatment.

The nonmetal content is the major fraction in the municipal waste. They range from paper, rags, plastics, vegetable matter, organic, and inorganic components, dust rubble, glass, and ceramics, hotel and hospital wastes, etc. Their characteristics also change greatly from city to city and also round the year in the same city. These solid wastes should be properly separated and the individual components recycled after suitable treatment. Unfortunately at present in most of the municipalities in India they are not physically separated and the useful components recovered and recycled. Also in most places the solid wastes are used as landfill. When the wastes contain large quantities of organic and inorganic soluble impurities, due to putrification and complex compound formations, during rainy seasons as well as in low-lying areas, these get percolated through soil and pollute the underground water table. Further, if some biowastes and hospital wastes are also present, pathogens are formed and give rise to all sorts of communicable diseases. In view of all these factors, it is essential to physically sort out the municipal wastes and treat them before they are disposed and recycled.

If the solid wastes are used for pyrolysis or incineration, care should be taken to see that subsequent air or water pollution problems will not arise. However, the solid wastes of Indian municipalities contain very low organic content as most of the paper, plastic, and rags get picked up by hawkers and hence the caloric value is very low for energy recovery.

2.10 SOCIOECONOMIC CONSIDERATIONS OF RESOURCE RECOVERY

The concepts and principles of recycling are well recognized. The role of recycling with regard to resource conservation and environmental protection is also known. However, in order to maximize the recycling of potential residues and to minimize the formation of wastes in all human activities, there should be an evaluation of strategies and policies based on an integrated approach regarding environmental, health, and socioeconomic consideration.

2.10.1 Economics of Resource-Recovery Systems

The subject of economics is currently very important in decisions regarding the feasibility of resource-recovery systems. However, the criteria used should not be limited to purely cost factors but should also reflect hidden costs that include social and manufacturing products. These hidden costs include social and economic costs of pollution, deprivation of recreational facilities, and dissipation of energy and resources.

There are two main economic areas that require determination in order to assess the profitability of a venture: costs and income. Costs are usually assessed as capital or fixed cost, i.e., the cost of running the plant. Income is a function of market size and realization, which are closely interrelated.

Viability quantities are the difference between income and expenditure, and take the factors of magnitude of investment, current commercial interest rates, and economic risk into account. Net present worth and discounted cash-flow rate of return (internal rate of return) are generally the best techniques for determining viability. Other methods include return on investment and payback time.

2.10.2 Capital-Cost Evaluation

Capital cost includes all construction and facility costs. Capital costs includes more than the cost of construction. Significant costs involved in the completion of a facility are as follows (Table 2.13):

- preliminary and final design;
- construction and system management: construction supervision, documentation, product marketing, operator training, acceptance testing;
- initial inventory: nonprocess equipment, furniture, scale house, laboratory, control center, tool crib, shops, store room, and initial spares;
- startup: 6–8 months to bring the plant to full capacity;
- interest to support the cash flow required to bring implementation;
- cost of the bond issue.

Capital cost is related to size, a given throughput in the gaseous phase is likely to need a physically larger plant than if the throughput were solid or liquid. Another factor is that solid materials tend to need more difficult, and hence more costly, handling systems.

Operating Systems: Operating or variable costs comprise all recurrent costs directly or indirectly incurred in manufacturing the product.

Table 2.13 Elements of capital-cost evaluation

Construction cost	Facility cost	System cost
Land site development & mobilization	Preliminary and final design Construction management	System development Engineering feasibility studies
Building/architectural	Laboratory equipment	Market surveys
Structural steel	Office furniture	RFP development
Foundations	Initial spares and supplies startup costs	Transfer stations
Process equipment		Fuel uses conversion
Plumbing	Testing programs	Working capital
HVAC	Testing and analyses	Capitalized interest expenses
Electrical	O & M manuals	Legal expenses
Escalation	Transportation equipment	Contingencies
Contractor OH and P	Maintenance equipment Contingencies interest during construction financial and legal fees	Special reserve funds financing costs access roads Utilities owners' administration cost

There are many constituent elements, all of which are conventionally estimated as a function of the following.

- raw materials
- labor
- energy
- selling price
- general administrative expenses

It is usual to express all individual operating costs as functions of one or more of the above cost elements. Averaging the results from a wide range of sources, an equation for determining the operating cost was developed.

$0 = 1.13\ R + 2.6\ L + 1.13\ E + 0.13\ 1$where,

0 = total operating cost

R = raw-material cost

L = direct labor cost

E = energy or utilities cost

1 = fixed capital cost

This represents a generalized expression for the total operating cost of a typical chemical process based on orthodox practices. Table 2.2 summarizes the comparative economics and feasibility of the main resource recovery and disposal options.

The cost of waste as raw materials in waste recovery is often zero. When the cost of alternative treatment is reduced or removed, a negative cost may be ascribed to waste. This may either be included on the credit side of the operating cost as income or included on the debit side as the cost of raw materials, if it may be adequately expressed in this way.

2.10.2.1 Marketing and Product Revenues

The test for economical viability is the ability to break even under public sector ownership. The test for competitiveness is whether the cost of disposal by resource recovery is less than that which could be achieved through possible options. Market size and realizations may be the most difficult areas to assess, particularly if an unusual or new product is just being introduced to the market. This factor is frequently most sensitive when evaluating a program that increases the importance of obtaining reliable and accurate forecasts.

The fraction of incoming refuse recovered as saleable material is determined by the expected efficiency of an operating plant and by the average expected composition of the incoming refuse (After, 1980). Byproduct revenues are based on expected annual recovery rates for each potentially recoverable resource and on the anticipated selling price for each material. This in turn is a judgment based on examination of analogous scrap prices quoted in trade journals, conversations with potential buyers, and freight charges over a likely distance.

It is important to point out the three sources of revenues for front-end recovery facilities. First, they can sell the recovered materials; second, they do not have to dispose of the recovered materials; and third, they can charge a fee for the service of preparing refuse for the landfill (Table 2.14).

2.10.2.2 Technical and Economic Risks

Profitability is related to risks and uncertainties involved in the venture as well as to cost or capital and the rate of inflation. For an established process, a return of 15–20% after tax is an acceptable return for a normal commercial venture. A waste-recovery process is likely to be considered more risky, and hence require a higher return to justify investment. The minimum acceptable rates of return is approximately equal to the cost of capital plus the rate of inflation plus an allowance for risk. Thus, this value varies from one locale to another.

Some of the risk areas associated with resource-recovery facilities are as follows:

Quantity of Waste: The plan for a recovery plant is economically justified when a set quantity of daily waste is ensured. Governments like waste to be provided on a "put or pay" basis for the amortized life of the plant, and therefore must know the amount of waste available for processing at startup and the amount likely to be available in the future. Because of the absence of any other reliable estimating basis the quantity of waste has been estimated by determining the average waste generated per capital and relating this to the size of the population and an estimate of population growth. The precautions required in using this estimate, as well as tradeoffs in using the rate measured on a given day or week, have been noted. Retrospective analysis of domestic waste collection shows that per capita generation has changed little in England; the figure increased only 10% by mass (50% by volume) over a 45-year period. There is much anecdotal evidence that the amount of waste delivered to a plant has been far below that planned or estimated from

Table 2.14 Comparative economics and feasibility of major resource and disposal options

Alternative	Feasibility	Net operating
Sanitary landfill	Institutional: there may be active citizen opposition to potential locations. Technical: depends on geological characteristics of the land Economic: decided savings in cost per ton if facility handles over 100 tons per day	\$1.50–\$8
Conventional incineration	Technical: feasible Economic: cannot economically meet new air pollution standards	\$8–\$15
Small incinerator	Technical: feasible Economic: varies with particular case	\$8–\$15
Steam generation from water-wall incinerators	Technical: several incinerators are in operation, only two are marketing the steam produced	\$4–\$10
Solid waste as fuel in utility or industrial boiler	Institutional: operator must contract with utility for sale of electricity	
	Technical: combustion in utility boiler as supplement to coal has been demonstrated in St. Louis Economic: practical feasibility depends on cooperation of local utility or user industry.	\$6–\$10
Pyrolysis: solid waste converted combustible gas and oil	Technical: has been demonstrated as 200-ton/day pilot plant Economic: transportability and quality of the fuel produced are primary factors. Ability to store and transport fuel others broad market application Technical: 1000-ton-per-day plant is in shakedown operation in Baltimore. Air pollution problems have been encountered. Economic: Markets for stream are limited	\$4–\$8

(Continued)

Table 2.14 (Continued)

Alternative	Feasibility	Net operating
Materials recovery: newsprint, corrugated and mixed office papers.	Technical: separate collection possibly with bailing, if corrugated paper is required to be recovered Economic: markets are variable, when paper prices are high, recovery can be profitable	\$4–\$12
Mixed paper fibers	Technical: technology has been demonstrated at 150-ton-per-day plant in Franklin, Ohio Economic: fiber quality from Franklin plants is low, suitable only for construction uses. Quality can be upgraded by further processing.	
Glass & aluminum	Technical: technology being developed Economic: market potential is adequate but system economics uncertain as yet.	\$7–\$13

national averages. A large difference between estimated and actual delivery can mean financial disaster for the facility.

Composition of Waste: Waste composition varies temporarily with the time of the week, the season of the year, the size of the community, and the region of the country. The composition is likely to change over the life of the recovery plant as technology, consumer preferences, and consumer affluence change. The amount of packaging material depends on economic affluence and food-distribution practices, including the availability of home refrigeration. The amount of food waste is indirectly proportional to these factors, and also changes with technical and economic changes in packaging and distribution.

Reliability of Equipment: The ability of all the equipment in the plant to operate to specification is often tenuous. Any recovery process will have a residue, and hence will require a landfill, which can also be used as the contingent disposal facility for public health maintenance. For material separation a 100% transfer facility is available for modification to be completed during the initial breakdown period.

Ability to Meet Product Specifications: There is little experience in this area and failure to meet product specifications can result in rejection and economic loss. Sometimes specifications for delivered steam or electricity cannot be met without the use of an auxiliary fuel, and this use must be provided for. Alternatively, imbalance between waste and steam supplies may necessitate discarding excess capacity during part of the year in order to have sufficient capacity to dispose of waste during the remainder of the year.

Marketability of Recovered Products: Secondary materials are marginal sources of raw materials. Thus, demand and price are subject to variations. Actions that increase the total demand for scraps of several grades are necessary.

Market surveys should be done as part of a feasibility study tailoring the product, particularly RDF, to suit the potential buyers; studying the products and marketing experiences of previous and related plants; sensitivity tests on product quantities and market values; and designing flexible plant capable of producing a variety of products.

Existing and Future Environmental Legislation: Managing the uncertainty of having to meet future and unforeseen environmental regulations may require additional investments for control technology in order for the plant to comply with the law. These are ordinary business risks for the private sector, but an unexpected and unwelcome expense for the public sector.

Plant Contractor/Operator Goes Out of Business: If the plant is operated for local authorities by a private contractor, the contract could well include some sort of bond situation to cover the costs of providing alternative disposal of processing routes. The cost of resource recovery is the tipping, which is determined by capital and operating costs of the recovery technology employed, less the revenue from the recovered products. For energy-recovery systems, the more that is invested in the system, the higher the revenue for the energy products. One common mistake is to compare the future cost of recovery with the current cost of disposal. The latter will increase with inflation and the increased difficulty of obtaining new sites, and the first cost of recovery is likely to be higher than landfill cost, but after a period of time, a break-even point is reached when the projected cost of recovery will be less than the projected cost of landfilling.

Thus, the community has to decide it will accept higher recovery costs (compared to an alternative disposal option) for the initial period, as an investment against breaking even and lowering the costs of recovery in future years.

In conclusion, the options for developing countries seem to be the following:

Materials recycling is an important part of the existing solid waste system in developing countries. Although scavenging is an unorganized operation that can occur at all stages of the system, resource-recovery schemes must recognize this and strive to incorporate it in the setup. Large-scale scavenging not only provides income to a small informal sector but also reduces the need for highly mechanized recovery systems. Controlling specific scavenging points in the system may be difficult, but a program by the municipality to organize scavengers into a recognized group and permit scavenging activities only at the dumpsites or processing centers may be a solution.

Most countries utilize landfilling as the most cost-effective option with the present economic situation. The possibility of recovering landfill methane gas from controlled tips should be investigated in future in relation to the local climatic conditions, technology and economics. Further land reclamation has been and will be an attractive option.

Another possibility is the use of RDFs as a substitute for coal. Western experience has shown RDF processing to be less expensive than mechanized material-recovery systems. Materials salvage as a preprocessing step recovers valuable metals and other materials that can be sold to secondary material dealers or to factories.

The more affluent Asian countries, e.g., Korea, Hong Kong, Singapore, and Taiwan, tend to favor incineration as a long-term option. But for countries where land cost and availability are not serious problems, salvage may be the major recovery method. Western mechanized plants are suitable when the refuse has western characteristics and the cost can be sustained. Otherwise, labor-intensive partly mechanized windrow systems with postfermentation treatment may offer a better prospect. Further, the Bhabha Atomic Research Centre (BARC) method mentioned earlier could be a good option for the future.

A major factor to be considered is the changing characteristics of solid waste in developing countries. Refuse is still largely organic in nature, but because of the increased economic activity in the region, there is a growing trend towards the use of paper and plastics in packaging. Hence, whatever processing options are chosen must be capable of handling the changing composition of waste. Since most resource-recovery options rely on a more or less constant refuse composition, salvaging of contraries or the addition of other waste materials (e.g., sewage sludge and agricultural wastes in composting and anaerobic digestion) may be necessary to maintain the process requirements.

An integrated approach for a total recovery system with salvage/composting as its core was developed by the nucleus group of Cal Recovery Systems, Inc. and embodies both thermal and biological methods of recovery as well as usable materials reclamation. It is modular in approach and flexible in application. Thus, the degree of mechanization can be varied to suit local conditions. Efficient and organized scavenging may be substituted for the more mechanized materials-reclamation units. However, there is at present no real example in developing countries along this direction.

Evaluation of Resource-Recovery Systems: The plea at this point is not to rush into energy recovery as the only available option because of the energy crisis and/or the partial failure of some recent material-recovery systems. The following subsections give a list of five criteria for the selection of solid waste—processing systems that engineers and community leaders may find helpful in selecting a total system concept to meet the needs of a given situation. The criteria are essentially independent, and though not fully analytical, will generally permit formulation of a figure of merit for each possible solution. Some measure of selection of the final alternative will thus be achieved.

Economic Viability: All things considered, the best system will be the one with the lowest net cost, assuming that the proposed system will meet other criteria. In some cases, sanitary landfilling may be the best solution on account of the availability of suitable land and the lack of strong markets for recovered materials. For some areas, comprehensive materials and recovery systems may be the only technically and politically viable solution. The more complex the system for resource recovery, the more expensive it will be to build and maintain. However, the better the quality of the resulting products, the higher the price they will command on the open market and the easier they will be to market. For very complex systems, marketing is critical and will help to dictate the type and quality of the products and hence the processes of necessity that will be used in the system. One should be prudent in installing expensive processes that produce high-quality products for which the market is nonexistent or long-term contracts are available.

Reliability of System: It is important and appropriate to consider recovery of valuable materials from the waste stream, but in addition to alluring the market, it is equally necessary to insist upon proven and reliable processes for materials handling. Municipal waste generation is a continuous process, and treatment and disposal must necessarily be reliable, continuous, and uninterrupted.

Flexibility: Numerous communities are located in regions with a widely varying climate, which produces significant changes in the composition and moisture content of the waste materials. The waste-processing system must be sufficiently flexible to handle such variations. More importantly, changes will occur as a result of changing consumer habits, legislation effecting waste-disposal practice, and the advent of new technology. Systems designed and built today should not be made obsolete or lose economic viability because of the failure to adapt to changing input or to take advantage of new technology. As far as possible, systems should be designed as front-end systems that can be supplemented by new technology for downstream materials processing when such additional equipment becomes available and reliable.

Energy Optimized: It is appropriate to maximize energy recovery and minimize energy use in materials processing, whether the fundamental purpose of the plant is materials recovery or energy production.

Environmental Acceptability: All new solid waste processes must consider the implicit and explicit environmental impact of their implementation, and those found inadequate must not be built. Like energy considerations, concern for the environment must be viewed in the larger context.

Systems Energy: The total amount of waste available for recovery is not the material amount usually estimated and reported officially, because not all of the waste can be collected and aggregated through processing. Thus, the amount of waste collected should not be used as a base for the amount of energy recoverable without correction for conversion and substitution efficiencies.

There is a tendency today to express new sources of fuel in terms of "layman's units" as "barrels of oil equivalent," which ignores the losses from the processing of waste to a fuel and from substitution of new fuel for conventional fuels. The new fuel may be used as a supplement to, or a substitute for, a commonly used fossil fuel, with or without passing through the conversion process. In a given application, the new fuel may operate with the same, greater, or less efficiency than the fuel it is replacing. Thus, the substitution efficiency is the amount of fuel in the new form that must be used to replace conventional fuel in specific applications. It is expressed as a ratio of the boiler efficiency of the new to the traditional fuel. The conversion equivalence is a way of expressing energy input and losses of a particular process.

It must be emphasized that there is no single-best method for the disposal of all wastes. The pattern will vary locally with the availability of land and the types and quantities of waste arising. In considering the different options it is necessary to choose a combination of methods most suitable for the particular situation and the general environment.

2.10.2.3 Recovery From Waste

1. Calculate total quantity of waste.
2. Analyze waste, for each load if necessary.
3. Calculate total quantity of each material contained in the waste.
4. Calculate total quantity of each material recoverable from the waste.
5. Ascertain or estimate value of each material in steps 3 and 4.
6. Multiply the total quantity of each material by the value. This gives an approximate maximum figure for the income to be achieved by selling that material (if not all the material may be recoverable, for example, because of dilution).
7. Rank the values (step 5) and the potential maximum incomes (step 6) in descending order.
8. Select the material that has the highest overall ranking of the two lists combined. This will ensure that the highest value material is investigated, which is a useful rule of thumb to follow, and appreciable and economical quantities, which is another useful rule of thumb.
9. Design a process to recover this material. At this stage only an outline flow diagram is required with some essential processing data. It is important to remember that not all the waste may need to be processed.
10. Estimate capital and operating costs.
11. Estimate income.
12. Calculate return on investment. This may be on a simple percentage return basis, or may enjoy a discounting method taking grants and taxes into account. The latter technique is a much more realistic way of assessing the profitability of a project.
13. If the return on the investment is sufficiently attractive, this is justification for a more detailed research investigation to confirm the results (Table 2.15).
14. The evaluation procedure (steps 8–13) should be repeated ideally for all materials but certainly for all materials worth more than 100 pounds per tonne. Below this rough guideline, profitable recovery becomes increasingly less likely as the value falls (Table 2.16).

Existing Municipal Solid Waste Management Practice (Fig. 2.21).

Table 2.15 Potential advantages and disadvantages of solid waste–processing systems and conditions that favor each

Alternative	Potential advantage	Potential disadvantage	Conditions thatfavor alternative
Materials-recovery systems	Less land required for solid waste disposal. High public acceptance. Lower disposal costs may result through sale of recovered materials and reduced landfilling requirements.	Technology for many operations still new, not fully proven. Required markets for recovered materials. High initial investment required for some techniques. Materials must meet specifications of purchaser.	Markets for sufficient quantities of the reclaimed materials are located nearby. Land available for sanitary landfilling is at a premium. Heavily populated area to ensure a large steady volume of solid waste to achieve economics of scale.
Energy-recovery systems	Landfill requirements can be reduced. Finding a site for an energy recovery plant may be easier than finding a site for a landfill or conventional incinerator. Total pollution is reduced when compared to a system that includes incinerator or for solid waste disposal and burning fossil fuels for energy. May be more economical than environmentally sound conventional incineration or remote sanitary landfilling. High public acceptance. As cost of fossil fuel rises, economics become more favorable.	Required markets for energy produced. Most systems will not accept all types of wastes.Needs of the energy market may dictate. Parameters of the system design, complex.Process requiring sophisticated management, needs relatively long construction between approval of funding, and full capacity operation. Technology for many operations is still new, not proven.	Heavily populated areas to ensure a large steady volume of solid waste to take advantage of economy of scale. Availability of a steady customer to generated energy to provide revenue. Desire or need for additional low sulfur fuel source.Land available for sanitary landfilling is at a premium.

Table 2.16 Comparison of resource-recovery operations

Process	Advantages	Disadvantages
Separation	Recovers many values such as metals, glass, and refuse-derived fuels (RDF); products relatively dealing with maximized resource conservation.	High-cost suitable outlets needed.
Composting	Refuse can be composted with sewage sludge attractive in areas where soil humus is depleted.	Expensive and leaves a proportion to be tipped; metal content of compost may limit its use.
Hydrology	Suitable for refuse with high paper content, producing sugars, protein, yeast, etc., for recovery.	Only theoretical exercises and small pilot projects on special trade wastes at present.
Incineration with heat recovery	Better method for district heating burn-out efficiencies can be expected with prepared fuel (RDF) than with unprepared refuse; commercially available plant can be developed to air-conditioning system; high-volume reduction of refuse sterile char.	Corrosion of boiler tubes at high steam temperatures; steam flow not sufficiently dependable to run power-plant auxiliary systems; high initial costs; slogging of heat exchange surface can give high cleaning costs and downtime; pollution problems.
Incineration with electricity generation.	Total electric power production package available. Good overall system efficiency possible revenue from material recovery high volume reduction of refuse sterile char	Serious technical problems with gas cleanup before turbine; new electrical generation equipment required; very high initial and running costs.
Pyrolysis to give oil and gas	Oil can be used in conventional boiler with minor modifications, existing power plant can be used higher-value products than incineration. Front or back and resource-recovery options may be included high volume reduction is refuse & sterile char. Overall disposal cist claimed to be less than landfill.	Technology unproven; problems with corrosiveness and storability of pyrolitic oil; high initial and operating costs: costly feed preparation; waste-water disposal problem.

(*Continued*)

Table 2.16 (Continued)

Process	**Advantages**	**Disadvantages**
Process	**Advantages**	**Disadvantages**
Pyrolysis to give gas and char/ slag gasification	Products have low-to-medium heating value; gas feed preparation not essential, although preferred existing power plant can be used; fairly high overall system efficiency; higher-value products than with incineration; fuel gas usable in most boiler types; technology more advanced than front- or back-end resource-recovery options may be included; gas may be employed as chemical feedback; high-volume reduction and sterile char.	Potential plugging of slag fuel gas not compatible with natural gas; without additional processing/ expenditure, storage of fuel not viable; high initial and operating cost; unproven viability; waste-water disposal problem; low heating value of gas necessities local use.
Solid fuel preparation as RDF	Gaining acceptance by manufacturers and users existing facilities can be used with minor modification to generate steam or electricity; revenue from other recovered materials; high overall system efficiency; relatively low costs; RDF improves storage and handling; largely proven technology; plant available commercially	Low bulk density of unprepared refuse makes storage difficult, potential increase in particulate loading and pollution. Densifying/palletizing equipment still presents problems, high costs and unproven viability.
Anaerobic digestion to give methane	Existing steam or electricity generation plan can be used revenue form other recovered materials possible product compatible with SNG after carbon dioxide commercially.	Sensitive to moisture and oxygen environment; very low overall system efficiency; product contaminated with carbon dioxide which requires separation; reaction rates very low, requiring large reactors and long residence times; residue disposal problem unless landfill is employed.
Fermentation to chemicals	Revenue from other recovered materials possible, technology well developed, high value products recovered.	Sensitive to contamination; high energy costs in purification from an aqueous base; high costs; residue disposal problem; viability doubtful.

Figure 2.21 Municipal solid waste management.

2.11 CASE STUDY

2.11.1 Integrated Municipal Solid Waste Management Strategy for Riyadh, Saudi Arabia

2.11.1.1 Waste Management Practice in a Compound

Riyadh, Saudi Arabia is in the process of rapid development of its metropolitan area, which presents major environmental challenges. The population growth is giving rise to a large increase in the generation of waste. Three-quarters of the Kingdom of Saudi Arabia's population currently resides in cities and 20% of the country's total population lives in Riyadh.

Furthermore, the population of Riyadh is expected to double, to approximately 8.3 million, by 2030.

The city produces approximately 8 million tonnes of waste per year, arising from municipal, commercial, industrial, and construction sources. The existing waste management infrastructure within the city is basic and the open landfill facilities are expected to reach full capacity in around 6–7 years. The ad hoc illegal disposal of waste is posing problems, with current regulation and enforcement achieving limited success.

To improve the situation, and to assist the ArRiyadh Development Authority, Ricardo-AEA has been appointed to deliver an integrated waste management strategy and implementation plans for the city of Riyadh, Saudi Arabia. The ambition is to treat all wastes as resources and maximize their reuse within the economy. Ricardo-AEA will thus develop an evidence-based sustainable waste management strategy for the city. The aim of the strategy is to reduce or mitigate the risk of adverse impacts of waste generation within the city through waste prevention, reuse, and recycling.

It is expected that implementing the strategy will divert increasing amounts of waste from landfills, whilst improving the resource efficiency of the commercial, industrial, and municipal sectors, offering economic opportunities from materials reuse and recycling, as well as from energy recovery.

To protect the environment and people's health and safety, the Saudi Cabinet has approved new regulations for the management of municipal solid waste in all cities and villages to protect the environment and ensure the safety of citizens and residents.

The regulations will ensure an integrated framework for the management of municipal solid waste. This includes waste separation, collection, transportation, storage, sorting, recycling, and processing. The Ministry of Municipal and Rural Affairs would be responsible for overseeing the tasks and responsibilities of the solid waste management system. The ministry has to develop appropriate programs to educate people about how to deal correctly with waste.

At present (2016), the waste is dumped in a landfill. Construction waste, demolition waste, and debris is collected (Fig. 2.22) in a separate trolley, which is dumped in far-off places.

Waste Management Practice: Neat and Clean Compound (Fig. 2.23)

A compound has around 280 houses with an average occupancy of 150 houses. One bin is placed in every house and between two houses, in the

Figure 2.22 (A) Construction/demolition sites maintained neatly; (B) carrying demolition/construction material; (C) dumping yard far from city.

(A)

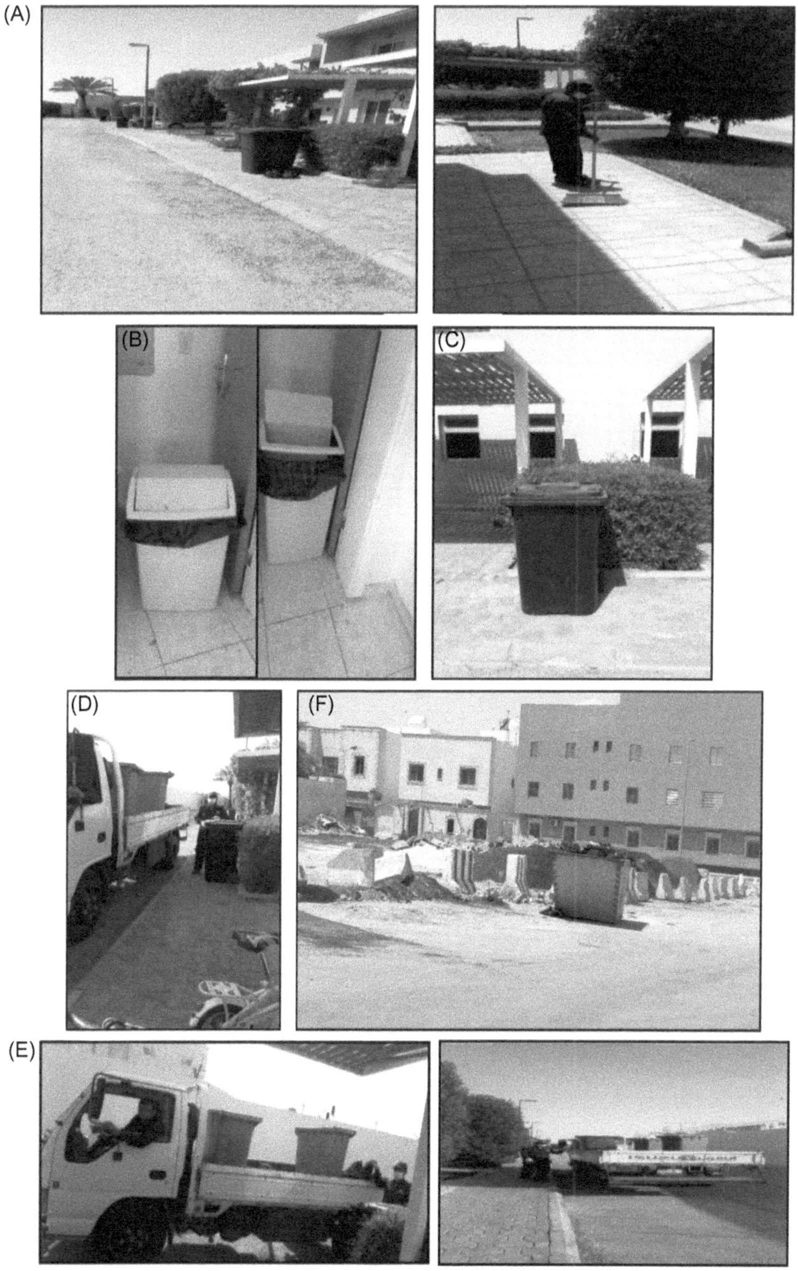

Figure 2.23 (A) The compound is always neat and clean. (B) Bins with black bags in residences. (C) Transfer bins in between two houses. (D) Dumping waste from household into transfer bin in between two houses. (E) Collection and carrying of waste to municipal bin. (F) Municipal bin.

walkway, one bigger bin is placed for the residents to transfer the household waste from their houses. The garbage is lifted from these bins, placed in the walkway by trucks, and disposed of in the municipal transfer bin, which is placed outside the compound. From the municipal transfer bin the garbage is lifted twice a day. Around six men are deployed for sweeping the compound. Sweeping is done as many times as needed. They perform their duty regularly to ensure a neat and clean compound. The greenery is maintained in the compound, the garden cuttings are separated and lifted on a payment basis and dumped in low-lying areas. Construction waste, demolition waste, and debris is collected in a separate trolley, which is lifted on a payment basis and dumped in far-off places. Thus, the compound is always maintained neatly.

2.12 LEGAL PROVISION

The Municipal Solid Wastes (Management and Handling) Rules, 2000, and amendments under the Environment (Protection) Act, 1986, governs the management of municipal solid waste. However, the legal provisions change from time to time depending on the minor amendments made in the Rules for handling and management of wastes.

ACKNOWLEDGMENT

This flow sheet is drawn based on the works of M/S Metcalf & Eddy Engineers, Inc. USA.

CHAPTER 3

Plastic Waste

3.1 INTRODUCTION TO PLASTIC WASTE

Plastics are ubiquitous materials and find applications in all parts of our life and economy. They are lightweight (energy saving) and low cost, and exhibit unique and versatile properties. They find use in agriculture, aviation, railways, telecommunication, building construction, electrical, electronics, medicine and health, automotive, packaging, thermal insulation, household, furniture, toys, and others. The sectoral application of plastics is depicted in Fig. 3.1.

The usage of plastic packaging's and products has increased multifold in the last one decade due to its low price and convenience. However, general public is not aware of its impact on the human and environment on littering or dumping. In India, approximately 12 million tonnes plastic products are consumed every year (2012), which is expected to rise further. It is also known that about 50–60% of its consumption is converted into waste. Main usage of plastics is in the form of carry bags, packaging films, wrapping materials, fluid containers, clothing, toys, household applications, industrial products, engineering applications, building materials, etc. It is true that conventional (petro-based) plastic waste is nonbiodegradable and remains on landscape for several years polluting the environment.

It is also well established that all types of plastic wastes cannot be recycled and therefore, it gets accumulated in open drains, low-lying areas, river banks, coastal areas, sea beaches, etc. Further, recycling of a virgin plastic product can be done 3–4 times only by mixing with virgin plastics granules. Therefore, after every recycling, its tensile strength and quality of plastic product gets deteriorated. Besides, recycled plastic materials are more harmful to the health and environment than the virgin products due to mixing of color, additives, stabilizers, flame retardants, etc. Looking at the seriousness of plastic waste disposal, Central Pollution Control Board, India has carried out a study in collaboration with Central Institute of Plastics Engineering and Technology, Ahmedabad, India on "Quantification and Characterization of Plastic waste generation in 60 Major Cities (2010–12)" in the country. It is reported that approximately, 3501 tonnes/day (TPD) of plastic waste (PW)

Solid and Hazardous Waste Management.
DOI: http://dx.doi.org/10.1016/B978-0-12-809734-2.00003-1

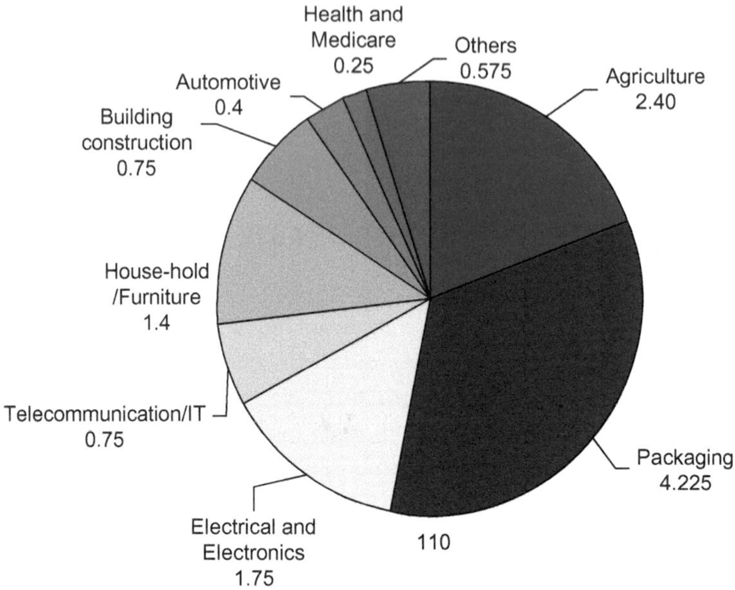

Figure 3.1 Sectoral application of plastics.

is generated in these cities. (STATUS OF IMPLEMENTATION OF PLASTIC WASTE MANAGEMENT (PWM), CENTRAL POLLUTION CONTROL BOARD (CPCB) Ministry of Environment and Forests, Government of India, November, 2015).

3.2 CHARACTERIZATION OF PLASTICS

There are mainly two types of plastics, namely commodity plastics and special plastics. The types of commodity plastics are polyethylene (PE), polypropylene (PP), polystyrene (PS), polyvinyl chloride (PVC), and polyethylene teraphthalate (PET), which are around 80% of plastics. The special plastics are engineered and specialty plastics are characterized as acrylonitrile butadiene styrene (ABS), styrene acrylonitrile (SAN), polyamide plastic pipe (PA), polybutalene teraphthalate polyester (PBT), polycarbonate plastic (PC), and polyurethane (PU), etc., which are around 20% of plastics.

3.3 TECHNOLOGIES FOR PLASTIC WASTE MANAGEMENT

There are various recycling/recovery options available for plastic waste. Material and energy can be recovered. Plastic waste can be reduced by using it in constructing asphalt roads.

Figure 3.2 (A) Mechanical recycling, (B) Products from mixed plastics.

3.3.1 Material Recovery

For material recovery, technological options are mechanical recycling, feedstock recycling, and biological recycling.

3.3.1.1 Mechanical Recycling

This is a conventional, most-value-added technology and has been most widely adopted. It is the most preferred and widely used recycling process (Fig 3.2) due to its cost effectiveness and ease of conversion to useful products of daily use. It needs homogeneous and clean input. The various products are niwar patti, plastic sacks, barsati film or shelter, box strapping, household utility (buckets, drainers, etc.). Mixed plastic waste can also be recycled to convert into useful products like lumber, garden furniture, etc. The various steps involved in recycling plastics are:

Collection → Identification → Sorting → Grinding → Washing → Drying → Separating → Agglomerating → Extruding/Compounding → Palletizing → Fabrication to product.

3.3.1.2 Feedstock Recycling

Selected segregated plastic waste is converted to basic monomer for repolymerizing into fresh plastic raw material. PET is an example.

3.3.2 Energy Recovery

Coprocessing in cement kilns, heat, and power generation are examples of energy recovery.

3.3.2.1 Coprocessing in Cement Kilns

In coprocessing in cement kilns, all types of mixed plastics waste can be used. In this process, segregation and cleaning is not required. Cleaner emissions are observed when compared to only coal. At 10% replacement rate, around 170 cement kilns in India can dispose of the entire plastics waste generated in the country today, thereby reducing the use of fossil fuel. The calorific values are:

Polyethylene:	46 MJ/kg
Polypropylene:	44 MJ/kg
Polyamide (nylons):	32 MJ/kg
PET:	22 MJ/kg
Coal:	29 MJ/kg

3.3.2.2 Heat and Power Generation

All types of mixed plastics waste can be used to convert the plastic waste into fuel. There is no need for elaborate cleaning. A catalytic depolymerization reaction takes place in the absence of oxygen and at temperatures below 350°C. It is a clean operation and there is no possibility of dioxin formation. The recovery claimed is close to 100% (liquid, gas, and solid). Around six units of electricity can be generated from 1 L of fuel.

3.3.3 Using Plastics Waste in Road Construction

All types of mixed plastics waste can be used to construct asphalt roads, including PE/PP/PS/expandable polystyrene (EPS). Multilayered plastics include 15% of total plastics waste for a 1-km-long and 7-ft-wide road, and 1 million tonnes of plastic waste is used with 9 million tonnes of bitumen in the bottom layer. Roads with seal coats require extra plastic waste.

3.3.4 Recycling, Recovery of Multilayer, and Laminated Plastic Packaging Waste

Any one of the following recycling/recovery process can scientifically and effectively be used for multilayer/laminated plastics like feedstock recycling, conversion to fuel, energy recovery, coprocessing in cement kilns, construction of asphalt road, mechanical recycling, and molding into compressed board/lumber.

3.3.5 Polyethylene Terephthalate

This is a versatile polymer useful in diverse applications, large production volume compared to commodity plastics, with a high-growth forecast and which has superior technical properties. Various PET uses are fibers (textiles), industrial fabrics, tarpaulins, tire cords, audio–video films, X-ray films, bottles and containers, films for packaging, and molded articles (structural foam, engineering items). Its process abilities are melt spinning, extrusion, injection molding, blow molding, thermoforming, and structural foams. Before going for chemical recycling processing or closed-loop recycling, a pretreatment process is necessary. The pretreatment process involves:

1. input control (check if within specification)
2. sorting non-PET (metal, other plastics, packaging, sorting color, e.g., transparent, blue, green)
3. prewashing whole bottles (removing labels), grinding flakes (e.g., down to <10 mm)
4. air separation (e.g., removing nylon barrier)
5. washing (removing glue)
6. swim/sink (removing polyolefins)
7. drying (e.g., 0.7% humidity)

After pretreatment, PET can be subjected to either chemical recycling or closed-loop recycling.

3.3.5.1 Chemical Recycling Process for PET

In the chemical recycling process the base monomers are regenerated, Tetradeconyl phorbol acetate (TPA) and Ethylene glycol (EG) are produced by hydrolysis and then are converted to oligomers through glycolysis. The glycolized product is used for value-added products and conversion to specialty chemical intermediates by aminolysis.

3.3.5.2 Closed-Loop Recycling for PET

Source-segregated waste of PET bottles is collected and treated by mechanical and chemical means to remove any contaminants. The cleaned PET can be melted and formed into pellets with the same physical properties as virgin PET. This reprocessed material can be used with virgin PET to make new bottles. Alternative recovery option is recovery of energy through a waste-to-energy plant. A life-cycle inventory (LCI) to compare the atmospheric emissions of key pollutants and their overall environmental impacts show the recycling option resulting in an overall reduction in the emission of each key pollutant and in the overall environmental impact. Chilton T, Burnley S, Nesaratnam S. A life cycle assessment of the closed loop recycling and thermal recovery of post consumer PET. Res Conserv Recy 2010; 54(12):1241–49.

Table 3.1 Effect of microwave power applied on the percentage of PET degradation at different times during microwave-assisted PET glycolysis

S.No.	Power (W)	Depolymerization time (min)		
		2	5	10
1.	50	35	127	100
2.	100	32	100	100
3.	150	100	100	100
4.	200	100	100	100

Source: Dimitris S, Glycolytic depolymerization of PET waste in a microwave reactor, J. Appl Polym. 2010; p. 3066–73.

Table 3.2 Terminology used in different types of plastics recycling and recovery definitions

S.No.	ASTM D5033	Equivalent ISO 15270 (draft)	Other equivalent terms
1.	Primary recycling	Mechanical recycling	Closed-loop recycling
2.	Secondary recycling	Mechanical recycling	Downgrading
3.	Tertiary recycling	Chemical recycling	Feedstock recycling

3.3.5.3 Glycolytic Depolymerization of PET Waste in a Microwave Reactor

Microwave irradiation as a heating technique offers many advantages over conventional heating, such as instantaneous and rapid heating with high specificity without contact with the material to be heated (Table 3.1).

3.3.5.4 PET to Polyurethanes

Higher modulus and elongation at break of PUs modified with the bis (2 hydroxy ethylene) terephthalate (BHETA) based PUs are useful raw material for coatings, adhesives, foams, and molded items. U. Ritter et al. Radiation modification of polyvinyl chloride nanocomposites with multi-walled carbon nano tubes. MATERIALWORSENSCHAFT UND WERKSTOFFTECHNIK, 2010; 41(8): 675–81 (Table 3.2).

3.4 LEGAL PROVISION

The Plastics Manufacture, Sale and Usage Rules, 1999, and Plastic Waste (Management and Handling) Rules, 2011, under Environment (Protection) Act, 1986, governs the management of plastic waste.

CHAPTER 4

Biomedical Waste

4.1 INTRODUCTION TO BIOMEDICAL WASTE

Biomedical waste means any waste that is generated during the diagnosis, treatment, or immunization of human beings or animals or in research activities pertaining thereto, or in the production or testing of biologicals, and including human anatomical waste, animal waste, microbiology and biotechnology waste, waste sharps, discarded medicines and cytotoxic drugs, soiled waste, solid waste (catheter, saline bottle, etc.), incineration ash, and chemical waste.

Waste is produced from all hospitals irrespective of their size. The waste from hospitals carries a higher potential for infection due to infectious waste and injury due to accidental needle prick. There is a risk of infection to medical staff, in- and outpatients, visitors, workers in support services, workers in waste-disposal facilities, and the general public. Around 20 bloodborne diseases can be transmitted if the waste is not managed properly. The staff of healthcare establishments, who are either in contact with the patient or the infectious waste generated, are continuously at risk during their working hours. The following types of occupational hazards occur/can occur:

- accidental cut or punctures from infected sharps such as, hypodermic needles, scalpels, knives, etc.;
- contact with infected material like pathological waste, used gloves, tubing, etc., especially from the operation theater (OT), bedding, and dress material of the patient or from the doctors (used during checkup, surgery, etc.);
- contact with stool, urine, blood, pus, etc. of the patients, especially during cleaning jobs.

Therefore, it is essential that adequate protection measures are to be provided against occupational health hazards. The administration of the healthcare establishment (the infection control officer in case of large ones) should have a detailed deliberation on this subject.

Solid and Hazardous Waste Management.
DOI: http://dx.doi.org/10.1016/B978-0-12-809734-2.00004-3

As a safety measure for the medical and paramedical staff the following instructions need to be notified and strictly adhered to:

- clear directives in the form of a notice to be displayed in all concerned areas;
- issuance of all protective clothes such as, gloves, aprons, masks, etc., without fail;
- sterilization of all equipment and issue of only properly sterilized equipment and tools, such as surgical tools to the medical personnel and maintenance of registers for this purpose;
- provision of disinfectant, soap, etc. of the right quality and clean towels/tissue paper;
- immunization to all medical care workers;
- provision of a wash area where they can take baths, if needed/desired;
- washing and disinfecting facility for the cleaning equipment and tools;
- regular medical checkup (at least half-yearly).

Environmentally sound management of biomedical waste is very important to protect health and the environment. The proper biomedical waste management will help to:

- control hospital-acquired infections;
- reduce HIV/AIDS, sepsis, and hepatitis transmission from dirty needles and other improperly cleaned or disposed medical items;
- control diseases passed to humans through insects, birds, rats, and other animals;
- prevent illegal repackaging and resale of contaminated needles;
- cut cycles of infection and avoid negative long-term health effects like cancer, from the environmental release of toxic substances.

As per *WHO norms*, healthcare waste includes all the waste generated by healthcare establishments, research facilities, and laboratories. In addition, it includes the waste originating from minor or scattered sources such as that produced in the course of healthcare undertaken in the home (dialysis, insulin injections, etc.).

4.2 EVOLUTION OF BIOMEDICAL WASTE MANAGEMENT

The establishment of a sustainable biomedical waste management system benefits from a national legal framework that regulates and organizes the different elements of waste management systems. Legislation usually places obligations and controls on what is permitted and prescribes sanctions on those that deviate from accepted practice. In reality, a law will remain ineffective if sources (financial, material, and knowledge) are not available in the healthcare sector to implement it and or if enforcement is weak.

The five guiding principles governing in waste-related laws are:

1. *Polluter pays principle*: This requires any waste producer to be made legally and financially responsible for the safe and environmentally sound disposal of their waste. The responsibility to ensure that the disposal of waste causes no environmental damage is placed upon each waste generator.
2. *Precautionary principle*: The rationale of the principle is that if the outcome of a potential risk is suspected to be serious, but may not be accurately known, it should be assumed that this risk is high. This has the effect of obliging healthcare waste generators to operate a good standard of waste collection and disposal, as well as provide health and safety training, protective equipment, and clothing for their staff.
3. *Duty of care principle*: This recognizes that any person managing or handling healthcare waste, or waste-related equipment, is morally responsible to take good care of the waste while it is under their responsibility.
4. *Proximity principle*: The philosophy behind this principle is that treatment and disposal of hazardous waste (including healthcare waste) should take place at the nearest convenient location to its place of generation, in order to minimize the risks to the general population. This does not necessarily mean treatment or disposal has to take place at each healthcare establishment, instead it could be done at a facility shared locally or at a regional or national location.
5. *An extension to proximity principle*: This is the expectation that every country should make arrangements to dispose of all wastes in an acceptable manner inside its own national borders, and the prior informed consent principle, also known as "cradle-to-grave" control, which introduces the concept that all parties involved in the generation, storage, transport, treatment, and disposal of hazardous wastes (including healthcare waste) should be licensed or registered to receive and handle the named categories of waste. In addition, only licensed organizations and sites are allowed to receive and handle these wastes. No hazardous wastes (including healthcare waste) should leave a place of waste generation until the subsequent parties (e.g., transport, treatment, and disposal operators and regulators) are informed that a waste consignment is ready to be moved.

The national legislation is the basis for biomedical waste management practices in India. It establishes control and permits for the disposal. The regulatory framework that governs the management of waste is as follows:

1. The Water (Prevention and Control of Pollution) Act, 1974 (for waste-water quality)
2. The Air (Prevention and Control of Pollution) Act, 1981 (for air quality)
3. The Environment (Protection) Act, 1986
4. Hazardous Wastes (Management, Handling and Transboundary Movement) Rules, 2008 (for hazardous waste)
5. The Biomedical Wastes (Management and Handling) Rules, 1998 (for healthcare waste)
6. The Municipal Solid Wastes (Management and Handling) Rules, 2000 (for domestic municipal waste)
7. Battery (Management and Handling) Rules, 2001 (for used batteries waste).

4.3 ORGANIZATION AND MANAGEMENT OF BIOMEDICAL WASTE

Biomedical waste management is a crucial process that starts from the point of generation and ends at the point of disposal. Policy on biomedical waste management needs to be evolved on the feasibility option and optimal sustainable treatment technologies in each individual hospital. The final disposal of biomedical waste in individual hospitals is discouraged as the hospitals are very much within the vicinity of residential areas, disposal of waste by incineration will lead to ambient air pollution. It is encouraged to have common biomedical waste treatment facilities (CBMWTFs) to treat all regional hospitals in one place away from the residential areas.

The Central Government has given guidelines for CBMWTFs to come up in each district of India to treat and dispose biomedical waste scientifically. Establishment of such facilities has reduced the burden of disposal for the biomedical waste generators. However, the generators have to segregate the waste properly as per biomedical waste management and handling rules in specific color-coded bins/bags and store in the central temporary storage room so that the common biomedical waste treatment facilitator can lift the waste from the individual hospital within 48 hours in dedicated vehicles and carry it to their facilities, away from residential areas, to treat and dispose. The path between the two points (cradle to grave) can be segmented schematically as categorization, quantification, segregation, handling, storage, treatment, destruction, and disposal of biomedical waste.

4.4 CHARACTERISTICS OF BIOMEDICAL WASTE

The characteristics of biomedical waste are noninfectious, infectious, hazardous, and cytotoxic. Noninfectious waste is the waste that is similar to household waste like wrappers, edibles, food, etc. Infectious waste includes pathological waste, surgical waste (body parts), sharps waste, items contaminated with blood and body fluids, etc. Cytotoxic and hazardous waste includes chemical waste, pharmaceutical waste, and discarded medicines.

4.5 CATEGORIZATION OF BIOMEDICAL WASTE

As per the Biomedical Waste (Management and Handling) Rules, 1998, biomedical waste has been categorized into 10 categories (Fig. 4.1), which are as follows:

- *Category 1*: Human anatomical waste (body parts, organs, human tissues, etc.)
- *Category 2*: Animal waste (animal tissues, organs, body parts, carcasses, bleeding parts, fluid, blood and experimental animals used in research, waste generated by veterinary hospitals, colleges, discharge from hospitals, animal houses)
- *Category 3*: Microbiology and biotechnology waste (wastes from laboratory cultures, stocks, or microorganisms; live or attenuated vaccines; human- and animal-cell culture used in research and infectious agents from research and industrial laboratories; wastes from production of biologicals, toxins, dishes, and devices used for transfer of cultures)
- *Category 4*: Waste sharps (needles, syringes, scalpels, blade, glass, etc. that may cause puncture and cuts, including both used and unused sharps)
- *Category 5*: Discarded medicines and cytotoxic drugs (waste comprising of outdated, contaminated, and discarded medicines)
- *Category 6*: Soiled waste (items contaminated with blood and body fluids including cotton, dressings, soiled plaster casts, lines, beddings, etc.)
- *Category 7*: Solid waste (waste generated from disposable items other than the waste sharps such as tubings, catheters, intravenous sets, etc.)

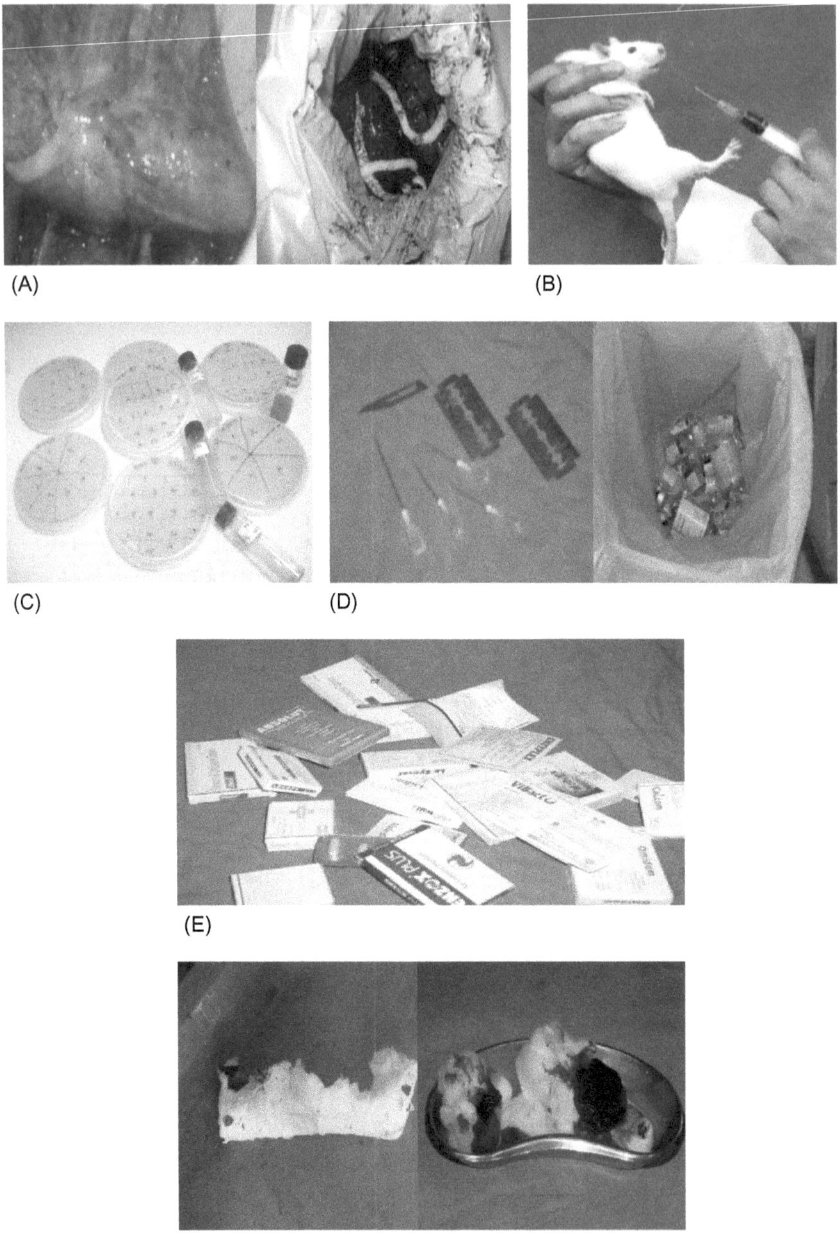

Fig. 4.1 Categorization of biomedical waste, (A) Category 1: Human anatomical waste, (B) Category 2: Animal waste, (C) Category 3: Microbiology and biotechnology waste, (D) Category 4: Waste sharps, (E) Category 5: Discarded medicine and cytotoxic drugs, (F) Category 6: Soiled waste, (G) Category 7: Solid waste, (H) Category 8: Liquid waste, (I) Category 9: Incineration ash, and (J) Chemical waste (liquid and solid).

Fig. 4.1 (Continued).

- *Category 8*: Liquid waste (waste generated from laboratory and washing, cleaning, housekeeping, and disinfecting activities)
- *Category 9*: Incineration ash (ash from incineration of any biomedical waste)
- *Category 10:* Chemical waste (chemicals used in production of biologicals, chemicals used in disinfection, as insecticides, etc.)

4.6 QUANTIFICATION OF BIOMEDICAL WASTE

A survey of all units in hospitals/healthcare establishments will help to identify and quantify biomedical waste generation. In almost all the units (outpatient, wards, OT, labor room, laboratory, intensive care units, etc.), waste is generated, the only difference being in category and quantity. As regards to the categorywise percentage of waste generation in any hospital, noninfectious waste is 80%, pathological and infectious waste 15%, sharps waste 1%, chemical or pharmaceutical waste 3%, and others 1%. To quantify the biomedical waste generation, a waste audit should be undertaken. The audit will give a clear picture of what type of waste, how much, and from where it is generated. This information will be helpful to do waste minimization, and determine items and equipment required for segregation and treatment of waste and their placement. To know how much and what type of waste is generated in each medical area, as a precursor to actual biomedical waste management planning, segregate the waste at the point of generation categorywise in specific color codes as per the Biomedical Waste (Management and Handling) Rules. Measure each category of bag by weighing them daily for 1 week and then average it to 1 month. If the segregation is not good then take the total weight, approximately 10–25% of which will be the infectious waste. The following steps will help in finding the waste generated quantitywise, categorywise and unitwise:

- Ascertain how many medical areas produce healthcare waste. List all the departments and study on its activities, production of waste and quantity.
- Find the composition of the waste in each place. Segregate waste categorywise, weigh it daily at least for one week and then average to monthly. The waste generated is not the same in all the areas producing waste.
- Along with the solid waste generation assessment, liquid waste assessment is also necessary.

Each establishment has to chalk out a program for a qualitative as well as quantitative survey of the waste generated depending on the medical activities and procedures followed by it. In order to assess the situation and to plan for biomedical waste management, the following have to be included (as applicable) in the survey as per the time frame indicated (Table 4.1).

Table 4.1 Areawise frequency of waste survey

Area/department/unit	Frequency of data collection
Wards (each one of them)	Each shift
Operation theater (OT)	Each operation/surgical procedure
Outpatients department (OPD)	Each shift
Intensive care unit (ICU)	Each shift
Emergency unit	Each shift
Dialysis unit	Each procedure
Radiation unit	Each procedure
Laboratories (pathological, biochemical)	Each shift
Pharmacy/chemist's dispensation unit	Once a day
Kitchen	After every meal
Administrative unit and central store	Once a day
Surrounding premises and garden	Once a day

The concerned medical establishment should constitute a team of experts and also involve personnel and workers from various departments (doctors, chemists, laboratory technicians, hospital engineers, nurses, cleaning supervisors/inspectors, cleaning staff, etc.). If expertise is not available, it may take the help of external experts in the field who can help them to carry out the survey work; then they can engage agencies that will carry out the whole work of biomedical waste management on contract as a package. The medical establishment has to earmark a suitable place for storing the biomedical waste temporarily in an enclosed space; depending upon the requirement, it can be a large room or a hall or at least a covered shade with proper fencing. Unauthorized entry to this space should be strictly restricted. It should be well lit. The place should be washed and disinfected daily and should be preferably dry and clean.

The waste generated by all the departments has to be collected according to the prevailing practices of collection but due care has to be taken to see that no portion of the total waste generated is left out of this survey. The waste so collected (except the liquid waste and incineration ash) has to be sorted out into the different categories according to the Schedule I of the Biomedical Waste (Management and Handling) Rules, 1998.

The liquid waste may be divided into two components: (1) liquid reagents/chemicals discarded, and (2) the cleaning and washing water that is channeled into the drain. The first component can be easily measured by a measuring cylinder or other suitable measuring device before discarding each time and keeping suitable records. The second component can be derived from the total water used in the hospital or by using the appropriate flow meter (Table 4.2).

Table 4.2 Categorywise survey of waste generation

Item as per Schedule I of Biomedical Waste (Management and Handling) Rules 1998	Wt. (kg) Shift I	Wt. (kg) Shift II	Wt. (kg) Shift III	Total wt. (kg)
Human anatomical waste				
Animal waste				
Microbiology and biotechnology				
Waste sharps				
Medicines and cytotoxic drugs				
Soiled waste				
Solid waste				
Chemical waste (solid)				
Incineration ash				
Liquid waste (L)				

The survey needs to be carried out at least for 3 days a week in continuation followed by a similar exercise for 4 weeks. The result is then compiled for both quantitative as well as qualitative data.

4.7 SEGREGATION, HANDLING, AND STORAGE OF BIOMEDICAL WASTE

Once the biomedical waste is generated the immediate step is segregation in specific color-coded bins/bags/containers. Human anatomical waste and animal waste is to be collected in yellow bins/bags, soiled waste in yellow or red bins/bags, sharp waste in white translucent puncture-proof containers, and solid waste (plastic) in either blue- or red-colored bin/bag. Bags and containers should be marked with the biohazard symbol. Outdated medicines and solid chemical waste is in black-colored bins/bag with the cytotoxic symbol on it (Fig. 4.2).

Segregation is a very important factor in waste management systems. The multiple choices of color codes for segregation mentioned above are dependent upon the treatment and disposal technology for various categories of wastes. The waste that goes to the incinerator or to deep burial should be collected in yellow bags or bins. The waste that is planned for autoclaving or microwaving or chemical treatment and finally to find its way in secured landfill or for recycling should be collected in red or blue bins or bags. Waste sharps, such as needles, blades, etc. that are for disinfection, destruction, or shredding should be collected in white puncture-proof translucent containers, which will be encapsulated or can go for recycling as final disposal. Chemical waste (solid), outdated medicines, and cytotoxic drugs that go for disposal in secured landfills should be collected in black bins or bags. All the bins and bags should have biohazard labels except on black-colored bins or bags on which the cytotoxic label is to be inserted. Maximizing segregation is very effective in reducing waste management costs, environmental impacts, and also complexity of management. The details of segregation of waste into specific color-coded bags or bins, as per treatment and disposal technology, are presented below (Figs. 4.3–4.7).

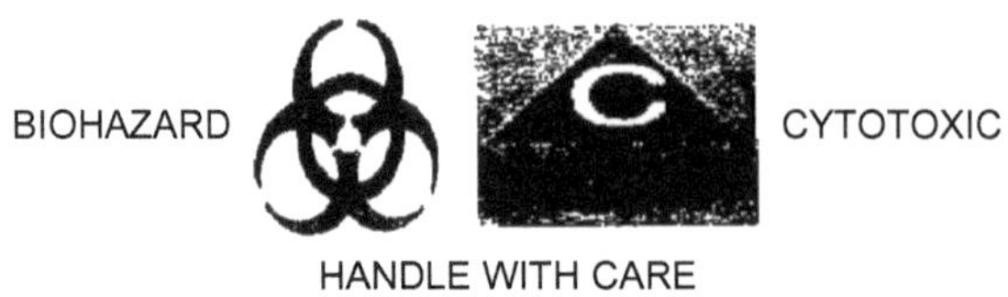

Figure 4.2 Biohazard and cytotoxic symbols.

Figure 4.3 Segregation of biomedical waste: Yellow bin (A) Category 1: Human and anatomical waste, (B) Category 2: Animal waste, (C) Category 3: Micro- and biotech waste, and (D) Category 6: Soiled waste.

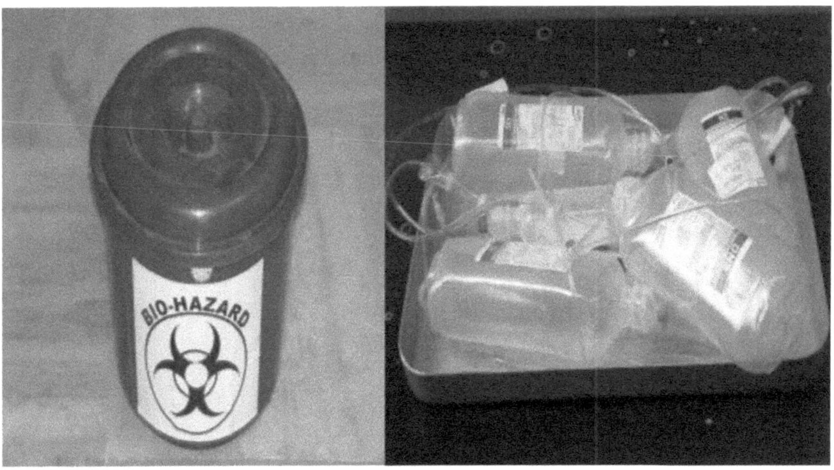

Figure 4.4 Blue bin. Category 7: Solid waste.

(A)

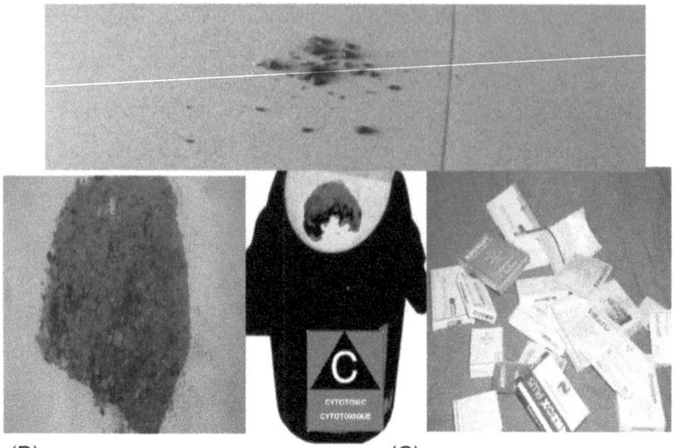

(B) (C)

Figure 4.5 Black bin. (C) Category 5: Discarded medicines and cytotoxic drugs, (B) Category 9: Incineration ash, and (A) Category 10: Chemical waste (solid).

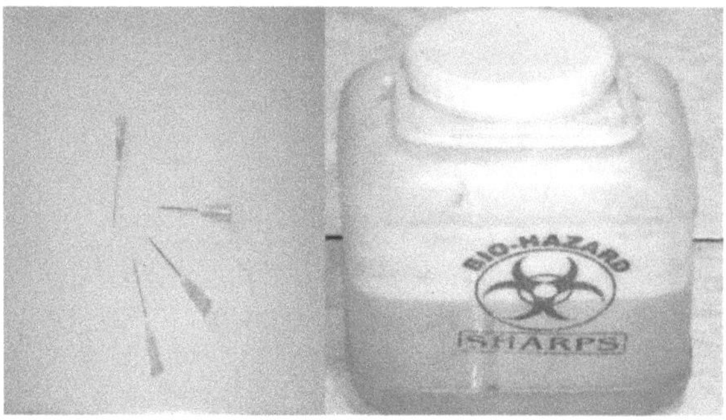

Figure 4.6 White bin. Category 4: Sharps waste.

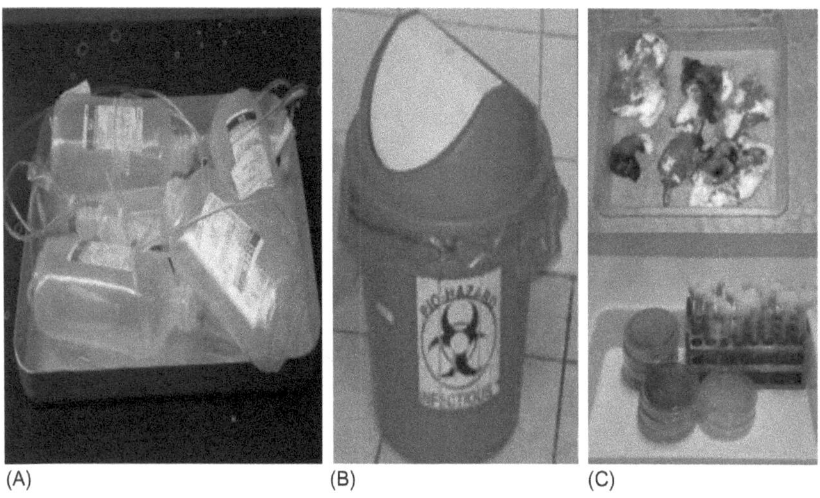

(A) (B) (C)

Figure 4.7 Red bin. (A) Category 7: Solid waste, (B) Category 3: Micro- and biotech waste, and (C) Category 6: Soiled waste.

Segregation of waste in specific colored bins depending on treatment and disposal technology

Yellow Plastic Bag: Nonchlorinated (Category 1: Human anatomical waste, Category 2: Animal waste, Category 3: Microbiology and Biotechnology waste, Category 6: Soiled waste) Treatment and disposal by incineration or deep burial

Blue Plastic Bag: (Category 7: Solid waste) Treatment and disposal by autoclaving or microwave or chemical treatment and destruction or shredding/recycling

White Translucent Puncture-Proof Container: (Category 4: Sharps Waste) Treatment and disposal by autoclaving or microwave or chemical treatment and destruction or shredding/encapsulation on secured landfill

Red Disinfected Container/Plastic Bag: (Category 3: Microbiology and Biotechnology Waste, Category 6: Soiled Waste, Category 7: Solid Waste). Treatment and disposal by autoclaving or microwave or chemical treatment—secured landfill and Category 7: Recycle

Black Plastic Bag (Category 5: Discarded Medicine and Cytotoxic Drugs), Category 9: Incineration Ash, Category 10: Chemical Waste (solid) Treatment and Disposal—Disposal in Secured Landfill

When the bags are three-quarters filled, they should be tied and lifted, and care should be taken that bags/containers are not be overloaded. The biomedical waste should be handled very carefully and stored in a central temporary storage room. The transport of biomedical waste from each unit to the central storage room should be in dedicated trolleys/vehicles. The trolleys should not be overloaded. While handling biomedical waste, there should be a barrier between biomedical waste and the body (self). Standard protective aids should be used to cover oneself.

The location of a central temporary storage room should be designated inside the premises. It should be kept under lock and key so that no unauthorized person can enter into the room. There should also not be any access to animals. The waste in the bags or containers should be stored in a central storage place in an area or room of a size appropriate to the quantities of waste produced and the frequency of collection. The recommendation for storage facilities within the hospital is that the storage area should have an impermeable, hard-standing floor with good drainage. It should be easy to clean and disinfect. There should be a water supply for cleaning purposes. The storage area should afford easy access for staff in charge of handling the waste. It should be possible to lock the store to prevent access by unauthorized persons. Easy access for waste-collection vehicles is essential. There should be protection from the sun.

The storage area should be inaccessible for animals, insects, and birds. There should be good lighting and at least passive ventilation. The storage area should not be situated in the proximity of fresh food stores or food-preparation areas. A supply of cleaning equipment, protective clothing, and waste bags or containers should be located conveniently close to the storage area. The cytotoxic waste should be stored separately from other healthcare waste in a designated secure location.

4.8 TREATMENT, DESTRUCTION, DISPOSAL OF BIOMEDICAL WASTE

After segregation, treatment of syringes, needles, and plastic waste is by mutilation and disinfection to avoid reuse and infection. The various treatment, destruction, and disposal methods for each category of waste are mentioned below.

Category 1: Human anatomical waste (human tissues, organs, body parts), Category 2: Animal waste (animal tissues, organs, body parts, bleeding parts, etc.), and Category 3: Microbiology and biotechnology waste (waste from lab, cultures, stocks or specimens human and animal cells, etc.) should be segregated in yellow-colored bins or bags and should be incinerated or subject to deep burial within 48 hours. The deep burial option is for towns where population is less than five lakh and in rural areas. There is no need to treat the waste before disposal.

Category 4: Waste sharps (needles, syringes, scalpels, blades, glass, etc., which may cause punctures and cuts, including both used and unused sharps) should be mutilated and disinfected. With regard to needles and syringes, after the injection is administered, the needles should be cut from the hub by a needle cutter, so that both the needle and the syringe become useless and cannot be reused. The cut needle gets segregated in the pot, which is fixed to the needle cutter. The cut syringe goes along the solid waste (plastic) stream, in the bucket with sieve, which has at least 1% sodium hypochlorite solution or any other equivalent chemical agent. The metal needle from the pot has to be transferred into the puncture-proof white translucent container having at least 1% sodium hypochlorite solution or any other equivalent chemical agent. It must be ensured that chemical treatment ensures disinfection. The disinfected needle can be encapsulated into municipal secured landfill or can be given to an authorized metal recycler. If autodisabled syringes are provided this prevents the reuse of nonsterile syringes, as they self-lock after a single use. The glass waste (vials, etc.) can be given to authorized glass recyclers.

Category 5: Discarded medicines and cytotoxic drugs (waste consisting of outdated, contaminated, and discarded medicines.), Category 9: Incineration ash and Category 10: Chemical solid waste (chemicals used in production of biological, chemicals used in disinfection, as insecticides, etc.), are either directly incinerated or after destruction are put in a secured landfill.

Category 6: Soiled waste (items contaminated with blood, and body fluids including cotton, dressings, soiled plaster casts, lines, beddings, other material contaminated with blood) is either incinerated or disinfected by autoclaving/microwaving and put it in a secured landfill.

Category 7: Solid waste (waste generated from disposable items other than waste sharps such as tubings, catheters, intravenous sets, etc.), destroy the plastic waste to ensure prevention of reuse and disinfect by keeping it in at least 1% sodium hypochlorite solution or any other equivalent chemical agent. It must be ensured that chemical treatment ensures disinfection. The solid waste (plastic waste) can be given to an authorized recycler only after disinfection and shredding.

Category 8: Liquid waste (waste generated from laboratories and washing, cleaning, housekeeping, and disinfection activities) and Category 10: Liquid chemical waste (chemicals used in the production of biologicals, chemicals used in disinfection, as insecticides, etc.) need to be treated to the standards prescribed in the Biomedical Waste (Management and Handling) Rules and flushed down the drains. The standard for liquid waste is as follows.

The various treatment and disposal options available to properly manage the biomedical waste is as given in Table 4.3.

As per the guidelines issued by Central Pollution Control Board disposal of biomedical waste by individual hospitals is discouraged and CBMWTFs are encouraged. The deep burial option is for rural areas where population is less than five lakh.

4.9 TECHNOLOGIES FOR BIOMEDICAL WASTE MANAGEMENT

Depending on the category of biomedical waste, various technologies have been evolved to treat, destroy, and dispose. Human anatomical waste, animal waste, microbiology and biotechnology waste, and soiled waste are

Table 4.3 Biomedical waste management: Treatment and disposal options

Category of waste	Treatment	Disposal
Soiled waste	Autoclave	Municipal landfill
	No treatment required	Incinerate
Solid waste	Mutilate and disinfect at source of generation	Recycling industry/ municipal landfill
Sharps	Mutilate and disinfect at source of generation	Recycle/encapsulate
Body parts	No treatment required	Incinerate/deep burial
Animal waste	No treatment required	Incinerate/deep burial
Discarded medicines	No treatment required	Dispose in secured landfill
Chemical solid waste	No treatment required	Dispose in secured landfill
Incineration ash	No treatment required	Dispose in secured landfill
General/domestic waste	–	Dispose in municipal bin

either incinerated or deep buried. The deep burial option is for rural areas where the population is less than five lakh.

4.9.1 Incinerator

Combustion efficiency (CE): at least 99.00%

$$\text{C.E.} = \frac{\%CO_2}{\%CO_2 + \%CO} \times 100$$

Temperature: primary chamber, 800±50°C.

Secondary chamber gas residence time: at least one second at 1050±50°C, with minimum 3% oxygen in the stack gas.

Emission standards are as follows. (Concentration mg/Nm^3 at 12% CO_2 correction). Minimum stack height: 30 m above ground.

Particulate matter: 150 units
Nitrogen oxides: 450 units
HCI: 50 and
Volatile organic compounds in ash shall not be more than 0.01%

Suitably designed pollution control devices should be installed/retrofitted with the incinerator to achieve the above emission limits, if necessary. Wastes to be incinerated shall not be chemically treated with any chlorinated disinfectants. Chlorinated plastics shall not be incinerated. Toxic metals in incineration ash shall be limited within the regulatory quantities as defined under the Hazardous Waste (Management and Handling) Rules, 2008. Only low-sulfur fuel like L.D.0./LS.H.S./diesel shall be used as fuel in the incinerator.

4.9.2 Deep Burial Pit

A pit or trench should be dug out about 2 m deep. It should be half filled with waste, then covered with lime within 50 cm of the surface, before filling the rest of pit with soil. It must be ensured that animals do not have any access to the burial site. Covers of galvanized iron/wire meshes may be used. On each occasion, when wastes are added to the pit, a layer of 10 cm of soil shall be added to cover the wastes. Burial must be performed under close and dedicated supervision. Pits should be distant from habitation so as to ensure that no contamination of ground water occurs. The area should not be prone to flooding or erosion. The institution shall maintain a record of all the pits for deep burial. Fencing of the deep burial pit has to be maintained. The deep burial site should be relatively impermeable and no shallow well should be close to the site.

The location of the deep burial site will be authorized by the prescribed authorities.

4.9.3 Autoclave

Microbiology and biotechnology waste, waste sharps, soiled waste, and solid waste (plastic waste) are subjected to the autoclave for disinfection. Temperature should be 121°C and pressure 15 pounds (psi) RT 60 m or 135°C and pressure 31 psi RT 45 minutes or 149°C and pressure of 52 psi RT 30 minutes. When operating a vacuum autoclave, medical waste shall be subjected to a minimum of one prevacuum pulse to purge the autoclave of all air. The waste shall be subjected to a temperature of 121°C and pressure of 15 psi RT 45 minutes, or 135°C and pressure of 31 psi RT 30 minutes.

4.9.4 Routine Test

A chemical indicator strip/tape that changes color when a certain temperature is reached can be used to verify that a specific temperature has been achieved. It may be necessary to use more than one strip over the waste package at different locations to ensure that the inner content of the package has been adequately autoclaved.

4.9.5 Validation Test (Spore Testing)

This test involves *Bacillus stearothermophilus* spores using vials or spore strips; with at least 1×10^4 spores/mm.

4.9.6 Sharp Pit

The sharp waste finds its way into the sharp pit. The detail of the sharp pit is as follows.

A pit is to be dug according to the requirement of the hospital. All the sides of the pit should be plastered with cement. A cylindrical metal pipe of 4-in. diameter or more is fixed at the ceiling of the pit. The opening of the metal pipe should have a locking facility. The sharps are deposited in this pit through the pipe from the puncture-proof translucent container after mutilating.

4.9.7 Shredder

After autoclaving, the plastic waste is subjected to shredding.

4.9.8 Secured Landfill

Landfilling shall be restricted to nonbiodegradable, inert waste, and other wastes that are not suitable either for recycling or for incinerating biological processing. Landfilling shall also be carried out for residues of waste-processing facilities. Landfilling of mixed waste shall be avoided unless the same is found unsuitable for waste processing.

Under unavoidable circumstances or until installation of alternate facilities, landfilling shall be done following proper norms.

4.9.9 Liquid Waste Disinfection

Liquid waste needs to be disinfected/treated and should conform to the following standards:

- pH: 6.3–9.0
- suspended solids: 100 mg/L
- oil and grease: 10 mg/L
- BOD: 30 mg/L
- COD: 250 mg/L
- bioassay test: 90% survival of fish after 96 hours in 100% effluent

These limits are applicable to those hospitals that are either connected with sewers without terminal sewage treatment plant or not connected to public sewers. For discharge into public sewers with terminal facilities, the general standards as notified under the Environment (Protection) Act, 1986, shall be applicable.

4.10 TOOLS AND EQUIPMENT REQUIRED FOR BIOMEDICAL WASTE MANAGEMENT

The tools and equipment required for managing biomedical waste are as follows:

- colored bins/bags (Fig. 4.8) (yellow, red, blue, and white translucent puncture-proof and black) having biohazard and cytotoxic symbols (for segregation of waste)
- big plastic container (for storing mutilated and disinfected plastic waste)
- needle cutter/needle burner (for destroying needles and syringes)
- autoclave/microwave (for disinfection)
- sodium hypochlorite solution (for disinfecting mutilated material)
- shredder (for cutting into pieces)
- incinerator (for incinerating waste)
- deep burial pit (for burial of waste)
- sharp pit (for keeping disinfected and mutilated sharps)
- scissors and knife (for destroying plastic waste)
- protective aids (for handling waste)

Figure 4.8 Specific colored bins and bags with biohazard and cytotoxic symbol.

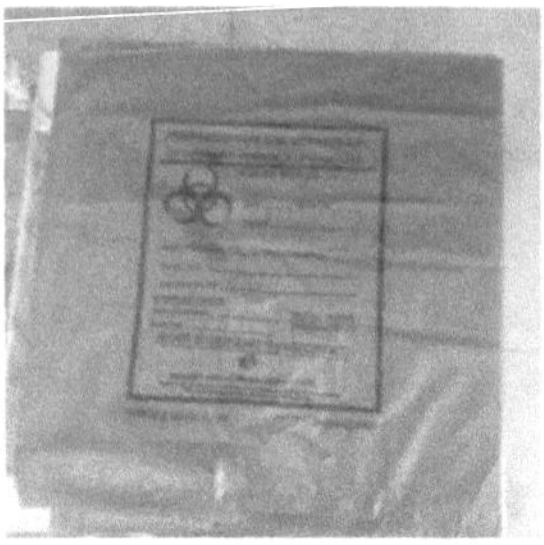

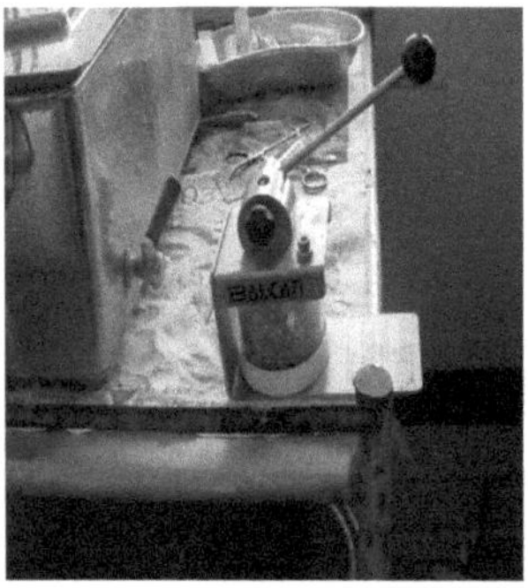

Manual needle cutter

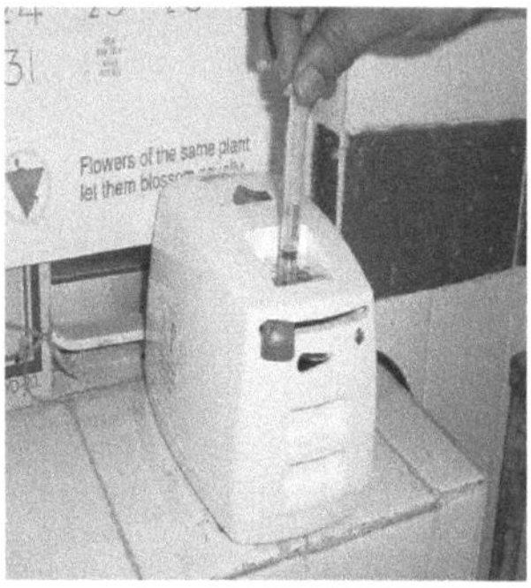

Electrical needle burner

Sharp pit

Autoclave

Shredder

Incinerator

Deep burial pit

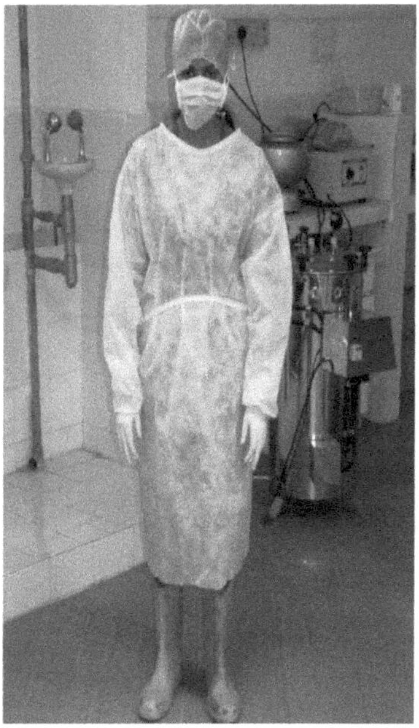

Protective aids

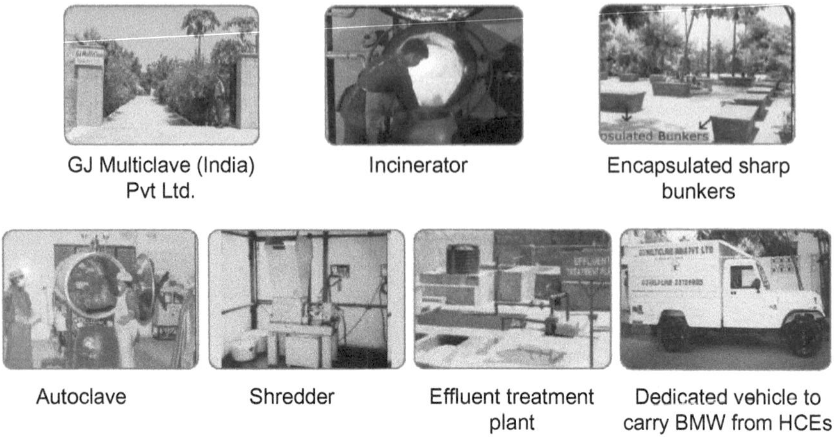

Figure 4.9 First biomedical waste treatment and management facility in Andhra Pradesh, G.J. Multiclave (India) Pvt Ltd.

4.11 CASE STUDY

4.11.1 A Typical Common Biomedical Waste Treatment Facility

G. J. Multiclave, a common biomedical waste treatment facility, caters to 790 hospitals with a 14,000-bed strength. Around 5 tonnes/day of biomedical waste is managed. They collect biomedical waste from hospitals using dedicated vehicles having the biohazard symbol on it. The waste in yellow goes for incineration, red goes for autoclave and shredding, and sharps waste in the white translucent bin is disinfected and encapsulated. The biomedical waste, which needs autoclaving, is 1.5 tonnes/day. Around 2.5–3 tonnes/day goes for incineration. The various facilities available at a common biomedical waste treatment facility are as follows: autoclave, incinerator, shredder, encapsulation structures, and waste-water treatment facility (Fig. 4.9).

4.11.2 Biomedical Waste Management Practice in the United States of America

Waste generated from hospitals comprises general waste, infectious waste, chemical, and hazardous waste. All the waste generated does not need to be regulated but the waste that needs to be taken care of is called regulated medical waste (RMW).

In New Jersey, the Department of Environmental Protection (DEP) and Department of Health and Senior Services (DHSS) formulated the Comprehensive Regulated Medical Waste Management Act N.J.S.A. 13:1E-48 et seq. (Comprehensive Act). The Department of Occupational Safety and Health Administration (OSHA) is one of the agencies in the Department of Labor that from time to time updates and also introduces new rules. OSHA has 26 regulations in different stages from a proposal to the final rule. The summary of the rules regarding hazardous material to comply with is that all work-related deaths must be reported within 8 hours of the accident and all work-related hospitalizations, amputations, or loss of an eye have to be reported within 24 hours of the accident in person at the nearest OSHA office. The other action items a healthcare facility must follow are to make sure all employees have received training on hazard communication standards, have a bloodborne pathogens exposure-control plan, ensure employees are able to identify biohazard waste and apply proper disposal procedures, and implement safe medical waste reduction practices. Periodically, the DHSS inspects RMW compliance at various healthcare facilities.

4.11.2.1 Regulated Medical Waste Management Procedure

The RMW management plan is designed to reduce the amount of waste, ensure regulatory compliance, strengthen infection control procedures, and optimize collection schedules and placement of equipment.

The generators of RMW are responsible for managing all the waste until it is destroyed, even once it leaves the facility. The procedures for proper processing, transportation, and ultimate disposal of RMW are taken and implemented from the Comprehensive Regulated Medical Waste Management Act (N.J.S.A. 13:1E-48) and the NJDEP Solid and Hazardous Waste Rules subchapter 3A: Regulated Medical Wastes (N.J. A.C. 7:26-3A).

4.11.2.2 Classification of Regulated Medical Waste

The New Jersey Regulated Medical Waste statute considers cultures and stocks of infectious agents and associated biological, human pathological waste, human blood and blood products, needles, syringes, and sharps, contaminated animal waste including carcasses, hospital isolation waste, and unused sharps as RMW.

The Regulated Medical Wastes subchapter 3A (N.J.A.C. 7:26-3A.6) defines RMW as solid waste that meets both the process definition and the classification definition, which is any solid waste generated from one of the following processes: the diagnosis, treatment or immunization of humans or animals; research pertaining to the diagnosis, treatment or immunization of humans or animals; or the production or testing of biologicals. Also, the items that are included in the process definition must also belong to one of the following seven classes of RMW:

- Class 1: Cultures and stocks (cultures and stocks of infectious agents and associated biologicals; cultures from medical or pathological labs; cultures and stocks of infectious agents from research labs; wastes from the production of biologicals; discarded live and attenuated vaccines; culture dishes and devices used to transfer, mix, or inoculate cultures).
- Class 2: Pathological wastes (human pathological wastes including tissues, organs, and other body parts and fluids that are removed during surgery or autopsy or other medical procedures; specimens of body fluids and their containers).
- Class 3: Human blood and blood products (liquid waste human blood; items saturated, dripping, or caked with human blood including serum, plasma, and other blood components) that were used or intended for use in either patient care, testing, and laboratory analysis, or the development of pharmaceuticals. Intravenous bags, soft plastic pipettes, and plastic blood vials are also included in this category.
- Class 4: Sharps (sharps that were used in animal or human patient care or treatment in medical research or industrial laboratories). Includes hypodermic needles, all syringes to which a needle can be attached (with or without the needle), Pasteur pipettes, scalpel blades, blood vials, carpules, needles with attached tubing, and broken or unbroken glassware (slides and cover slips) that were in contact with infectious agents.
- Class 5: Animal waste (contaminated animal carcasses, body parts, and bedding of animals that were known to have been exposed to infectious agents during research, production of biologicals, or testing of pharmaceuticals).
- Class 6: Isolation waste (biological waste and discarded materials contaminated with blood, excretions, exudates, or secretions from humans or animals that are isolated to protect others from certain highly communicable diseases).
- Class 7: Unused sharps (unused, discarded sharps that were intended to be used. Includes hypodermic needles, suture needles, syringes, and scalpel blades).

4.11.2.3 Segregation of Regulated Medical Waste

The RMW generated is segregated and grouped into three categories instead of seven categories mentioned earlier, the three categories are:

Category 1. Sharps (both Class 4 and Class 7)

Category 2. Fluids (greater than 20 cc)

Category 3. Other RMW

Each type of RMW is collected in separate inner containers that will ultimately be placed into the outer cardboard container (i.e., needles, glass cover slips, scalpel blades, and syringes are collected in a sharps container; culture transfer devices, blood-soaked items, and other paper- or cloth-related items are collected in autoclave bags or red RMW liner bags.). Hypodermic needles or syringes are not chopped, bent, broken, or otherwise destroyed before discarding them into the sharps container.

4.11.2.4 Treatment of Regulated Medical Waste

Generally, it is not necessary to treat RMW before placing it in the outer cardboard container for ultimate disposal for collection by the RMW vendor.

4.11.2.5 Storage of Regulated Medical Waste

Outer containers are stored in a secure area protected for the elements, vandalism, insects, and rodents. No unauthorized personnel can enter into this area. When storing containers their labels face outward so that they can be easily seen. Containers are sealed securely to prevent spillage or the leaking of vapors. Liquids (e.g., blood) are put into containers that are packaged with a sufficient amount of surrounding absorbent material to prevent leakage. Volumes of liquid may not exceed 20 cc per individual container. NJDEP Solid and Hazardous Waste Rules subchapter 3A: Regulated Medical Wastes (N.J.A.C. 7:26-3A) allows RMW to be stored onsite for up to 1 year. In order to comply with this subchapter, RMW generators dispose of RMW containers on a yearly basis, even if RMW containers are not full. However, it is recommended for frequent disposal of RMW boxes.

4.11.2.6 Packaging, Labeling, and Marking Requirements of Regulated Medical Waste

The generators package all RMW before the RMW vendor can remove it. All needles, syringes, scalpels, and any sharp objects are packaged in the appropriate puncture-resistant sharps container. Unbroken as well as broken glass is packaged to prevent puncture of the outer RMW container. All other items are packaged in autoclave bags or other appropriate inner containers. These items are then packaged in appropriate medical waste boxes before removal.

Each package of RMW is marked by generators according to the following labeling and marking requirements before it is transported offsite by the RMW vendor. The outermost surface of each cardboard box prepared for shipment is labeled with a special water-resistant identification label called the medical waste outer container label, which provides the information on the campus, building, and room where waste was generated. If these labels are unavailable, the required information is written directly on the outside of the box. Only indelible or waterproof ink or marker fluid is used to complete this label.

The generator also labels the inner containers only with a special water-resistant identification label called the medical waste inner container label, which provides information on the campus, building, room, phone number, and contact person for the location where the waste is generated. If these labels are unavailable, the required information is written directly on the inner container. Only indelible or waterproof ink or marker fluid is used to fill out this label.

4.11.2.7 Transportation of Regulated Medical Waste

The RMW is transported by vendor to an approved RMW incinerator for offsite incineration. The NJ medical waste tracking form is used to ensure proper transportation of RMW to an appropriate disposal site. The RMW vendor will fill out the tracking form. The generator checks items 1–14 on the tracking form for the purposes of verifying the accuracy of the information listed. After a thorough review of items 1–14, the generator then signs item 15 off the tracking form. After the RMW transporter has also signed off item 16, a copy of the tracking form is given to the generator. After the RMW is received by the disposal facility, a disposal facility representative will sign in item 22. A copy is mailed back to the generator. Both copies of the tracking form are kept by the generator at the generation site for at least three years from the date the waste is accepted by the RMW transporter. The destination facility must send copy 1 to the generator within 15 days of receipt of the tracking form from the RMW hauler. The RMW vendor will pick up waste depending on the date of request, allowing up to 16 days for the RMW transporter to remove the waste. The RMW vendor will not pick up the waste without a representative of the RMW generator being present to sign the RMW tracking form.

4.11.3 Alternative Regulated Medical Waste Management Technologies

The following waste disposal technologies are alternatives to incineration that have been authorized by NJDEP and the Department of Health and Senior Services to operate in New Jersey. This technology is approved for treatment only and therefore all medical waste processed must be managed as RMW in accordance with N.J.A.C. 7:26-3A unless the sterilizer is used in conjunction with a shredder/grinder approved by NJDEP that destroys the waste.

1. *Steam sterilization and shredding*: Air is evacuated from the sterilization chamber and steam is injected into the chamber. The treated material is shredded and ground.
2. *Chemical disinfection and mechanical shredding*: A chemical disinfectant is mixed with the waste and then the material is shredded and ground in a mechanical grinder or hammer-mill chamber.

   NaOCL applied to RMW is then dropped into shredder. After shredding, more chemical and water are applied, and then solid and liquid waste are separated.
3. *Microwave and shredding*: Waste is shredded and moistened with steam. The material is then microwaved in a treatment chamber and shredded, and then ground in a particulizer.
4. *Steam sterilization*: RMW is steam sterilized. High-vacuum treatment boils off and condenses liquid. RMW is dried and cooled to below 170°F (approved for treatment only. Processed medical waste must still be managed as RMW).

4.11.4 Management of Chemotherapy Hazardous Waste in Maryland

Chemotherapy is a medical procedures generating hazardous waste. The waste may not be infectious, but it is highly toxic and corrosive. If disposed of improperly, these wastes may cause irreparable damage to natural resources. The detailed practice of chemotherapy hazardous waste management in Maryland is presented below.

The Resource Conservation and Recovery Act (RCRA) was enacted by the U.S. Environmental Protection Agency (EPA) to mandate procedures for disposal of wastes that are deemed hazardous. There are four lists of substances that are specifically listed as hazardous under the RCRA: F-list (nonspecific source wastes), K-list (source-specific wastes), P-list, and U-list (discarded chemical products). When it comes to chemotherapy, it is most likely to deal with P-list and U-list substances such as cyclophosphamide, daunomycin, or melphalan.

The characteristics of hazardous waste include toxicity, ignitability, corrosivity, and reactivity. It is the medical waste generator's responsibility to identify these characteristic hazardous wastes and manage them accordingly. All hazardous waste needs to be separated, properly packaged, and labeled and disposed of in a black waste container as per EPA's mandate.

Chemotherapy waste is described as trace and bulk; although these are not official EPA designations, these terms are widely used in the industry.

Trace chemotherapy waste usually includes vials, bags, IV tubes, and other items that used to contain chemotherapy drugs, but now qualify as RCRA empty. A container that used to hold a P-listed substance is RCRA empty only after it is triple-rinsed. A container that used to hold a U-listed substance is RCRA empty when there is less than 3% of the former volume left. If the applicable criteria are met, then this waste can be considered trace waste and can be managed as RMW and disposed of in yellow containers. Otherwise, it should be managed as hazardous waste.

Bulk chemotherapy waste usually includes items that used to contain chemotherapy agents and that don't qualify as RCRA empty. Other types of bulk waste include materials used to clean up chemo spills and visibly contaminated personal protective equipment (PPE). The bags of chemotherapy drugs that were unused but are out of date are not necessarily hazardous waste. A substance becomes waste when it can no longer be reused, and expired drugs can sometimes be returned to the manufacturer through a take-back program.

4.11.5 Controlled Substances

This includes morphine or hydrocodone (powerful painkillers) and other narcotic categories of pain medicines and certain sleeping medications like diazepam, lorazepam from the benzodiazipine category, or medications that are considered highly regulated or controlled substances. When disposing of these and other pharmaceutical waste, the healthcare facility to check whether these drugs are listed as controlled substances as per Drug Enforcement Administration (DEA) schedule. The DEA has not specified any particular packaging to be used for collecting these controlled substances but it does have specific regulations in regard to their disposal. If the facility works with controlled substances on a regular basis then it is to be registered with the DEA. Disposal procedures may differ between DEA registrants.

Note: The federal regulations may be common for across all states but different states have different rules and regulations of their own apart from federal government rules based on their resources and may have state-specific waste requirements.

4.12 LEGAL PROVISION

The Biomedical Waste Environment (Protection) Act, 1986, governs the management of biomedical waste. (Management and Handling) Rules, 1998, and amendments under. However, the legal provisions change from time to time depending on the minor amendments made in the Rules for handling and management of wastes.

CHAPTER 5

Hazardous Waste

5.1 INTRODUCTION TO HAZARDOUS WASTE

Hazardous waste can be defined as any waste that because of its physical, chemical, biological, or infectious properties has the potential to cause irreparable damage and incapacitative illness to human health and/or the environment. It is a discarded substance whose chemical nature makes it potentially dangerous to people. Hazardous wastes are corrosive, toxic, flammable, and reactive substances that present a threat to public health, safety and the environment.

Today, more than 13,000 licensed industries generate about 4.4 million metric tonnes of hazardous waste every year. This does not include small-scale businesses such as backyard smelters, etc. According to the Ministry of Environment and Forests (MoEF), Government of India, about 80% of the hazardous waste is generated in Maharashtra, Gujarat, Tamil Nadu, Karnataka, and Andhra Pradesh. Unsound practices have caused widespread degradation of the environment and adverse health impacts on industrial workers and communities. Economic liberalization policies in the past 40 years or so have led to rapid growth in Indian industries. The production of petrochemicals, pesticides, pharmaceuticals, textiles, dyes, fertilizers, leather products, paint, and chlor-alkali has grown significantly. These industries produce wastes containing heavy metals, cyanides, pesticides, complex aromatic compounds (such as polychlorinated biphenyls), and other toxics. Several toxic waste hot spots such as the industrial belt of Vapi and Vadodara in Gujarat, Thane and Belapur in Maharashtra, and Patancheru and Bollarm in Andhra Pradesh developed in this period. The adverse impacts of hazardous waste on air, water, soil, flora, and fauna, and the global environmental episodes like the Minamata Bay incident in Japan, Love Canal incident in the United States, PCB leak in to the Mediterranean Sea, etc., have resulted in strict legislations across the world. India woke up to the dangerous realities of industrial hazards after the Bhopal disaster in 1984. The government enacted the Environment (Protection) Act, 1986, and under this act, the Hazardous Waste (Management and Handling) Rules, 1989, were formulated. Each industry generating hazardous waste should obtain authorization

Solid and Hazardous Waste Management.
DOI: http://dx.doi.org/10.1016/B978-0-12-809734-2.00005-5

from its respective state pollution control boards and pollution control committees.

The hazardous effects are imposed both intrinsically and/or extrinsically. The term "toxic" commonly refers to poisonous substances that cause serious injury to humans and animals by interfering with normal body physiology. Effects of certain chemicals on human health are as follows:

- Mercury: irreversible neurological damage
- Cadmium: renal dysfunction when renal cortex cadmium is around 200 mg/kg, acute respiratory effects at higher doses
- Lead: neurological impairment, gastrointestinal effects, and renal disease
- Arsenic: lung and other cancers, skin and mucous membrane disorders, neurological effects, hematological effects, visual impairment, etc.
- Formaldehyde: carcinogenic potential, skin and eye irritant, respiratory tract irritation
- Dioxins: potent carcinogen

5.2 MANAGEMENT OF HAZARDOUS WASTE

The industrial hazardous waste management starts from the collection of the waste at the industry (generator of the waste) and ends at the disposal by means of secured landfill or incineration. It is always preferable to look for the reuse/recycle possibilities of wastes as a better system of waste management. Waste characterization is the first step right from the identification of a waste as hazardous to find out the safe disposal methodology. While handling and management of hazardous wastes all the activities such as characterization, collection, transport, reception, treatment, storage, and disposal, including the associated activities like waste stabilization studies, reuse/recycle options, and alternate destruction technologies, have to be addressed.

5.3 CHARACTERIZATION OF HAZARDOUS WASTE

The Hazardous Waste Management and Handling Rules under Environment Protection Act, 1986, clearly guide in the identification of hazardous wastes based on Schedule 1 (through industrial processes) Schedule 2 (concentration of constituents) and Schedule 3, (the characteristics of the waste). The waste generated during industrial activities should be tested to find the nature of the waste, i.e., whether it is hazardous waste or not. The important characteristics of hazardous waste are ignitability, corrosivity, reactivity, and toxicity.

5.3.1 Ignitability

A waste is categorized as ignitable hazardous waste if it is:

- a liquid, other than an aqueous solution, containing <24% alcohol by volume, and it has a flash point <60°C;
- a liquid and is capable, under standard temperature and pressure, of causing fire through friction, absorption of moisture, or spontaneous chemical changes and when ignited, burns so vigorously and persistently that it creates a hazard;
- an ignitable compressed gas;
- an oxidizer that yields oxygen readily to stimulate the combustion of organic matter (e.g., chlorate, permanganate, inorganic peroxide, or nitrate)
- ignitability can be determined by Pensky–Martens closed cup method and Setaflash closed cup method. The details are as follows:
  - *Pensky–Martens Closed Cup Method*: The sample is heated at a slow, constant rate with continual stirring. A small flame is directed into the cup at regular intervals with simultaneous interruption of stirring. The flash point is the lowest temperature at which application of the test flame ignites the vapor above the sample.
  - *Setaflash Closed Cup Method*: By means of a syringe, 2 mL of sample is introduced through a leak-proof entry port into the tightly closed Setaflash tester or directly into the cup, which has been brought to within 30°C below the expected flash point. After 1 minute a test flame is applied inside the cup and note is taken as to whether the test samples flashes.

5.3.2 Corrosivity

A waste is designated as corrosive hazardous waste if it is

- Aqueous and has a pH less than or equal to 2 or greater than or equal to 12.5.
- A liquid and corrodes steel at the rate >6.35 mm/year at a test temperature of 55°C.

Corrosivity can be determined by the test, which involves coupons of steel (SAE Type 1020) subjected to the test waste that is to be evaluated, and by measuring the degree to which the coupon has been dissolved, one can determine the corrosivity of the waste.

5.3.3 Reactivity

A solid waste exhibits the characteristic of reactivity if a representative sample of the waste has any of the following properties:

- It is normally unstable and readily undergoes violent change.
- It reacts violently with water.
- It forms potentially explosive mixtures with water.
- When mixed with water it generates toxic gases, vapors or fumes in a quantity sufficient to present a danger to human health or to the environment.
- It is a cyanide- or sulfide-bearing waste that when exposed to pH conditions between 2 and 12.5, can generate toxic gases, vapors, or fumes in a quantity sufficient to present a danger to human health or environment.
- It is capable of detonation or explosive reaction if it is subjected to a strong initiating source or if heated under confinement.
- It is readily capable of detonation or explosive decomposition or reaction at standard temperature and pressure.
- It is a forbidden explosive.
- A solid waste that exhibits the characteristic of reactivity, but is not listed as a hazardous waste.

5.3.4 Toxicity

The method of toxicity characteristics leaching procedures (TCLP) is applicable to the determination of mobility of metals and semivolatile organic compounds in soils. The complete evaluation of this would require two extractions, one for volatile and semivolatile compounds, and the other for metals.

The TCLP test consists of five steps, namely separation procedure, particle-size reduction, extraction of solid material, final separation of the extraction from the remaining solid, and testing/analysis of TCLP extract. Apparati required for TCLP test are the agitation apparatus and extraction apparatus, the details of which are as follows (Fig. 5.1):

- *The agitation apparatus* must be capable of rotating the extraction vessel in an end-over-end fashion at 30 $\pm$ 2 rpm. The criteria is to prevent stratification of the sample and extraction fluid ensuring that all sample

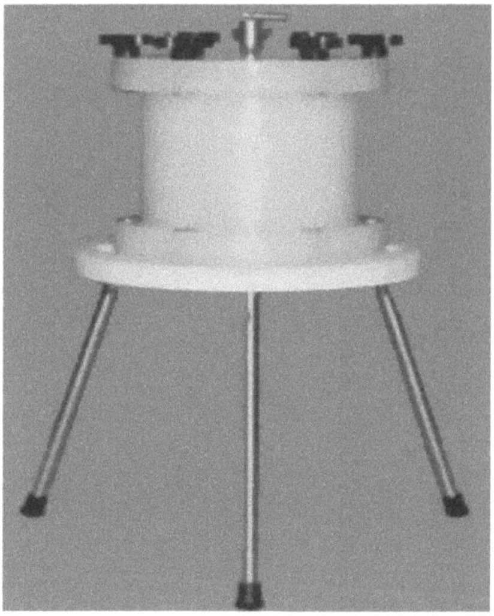

Figure 5.1 Pressure filtration unit (for the initial solid/liquid phase separation).

surfaces are continuously brought into contact with well-mixed extraction fluid (Fig. 5.3).

- *The extraction apparatus* is a zero head-space extraction vessel. This is for use when the waste is being tested for the mobility of volatile analyses. The zero head extraction allows for liquid/solid separation within the device and allows for initial liquid/solid separation, extraction, and final extract filtration without opening the vessel with an internal volume of 500–600 ml and accommodate a 90–100 mm filter (Figs. 5.2 and 5.14).

5.3.5 Tests for Hazardous Wastes: List of Parameters

- Physicochemical characteristics are specific gravity/density, leachable metals and semivolatile organics, moisture content, soil pH, electrical conductivity, and oxidizable organic carbon.
- Nutrients such as total Kjeldahl nitrogen, ammonium nitrogen, nitrite nitrogen, nitrate nitrogen, and phosphate.
- Metals such as metals by microwave digestion, aqua-regia digestible metals, mercury, low-temperature acid-digestible metals, EDTA-extractable metals, exchangeable cations and cation-exchange capacity .
- Inorganics such as water-soluble chloride, boron, bromide, total cyanide, soluble cyanide, fluoride, total sulfur, sulfate, sulfide, and elemental sulfur.

Figure 5.2 Zero head-space extraction vessel.

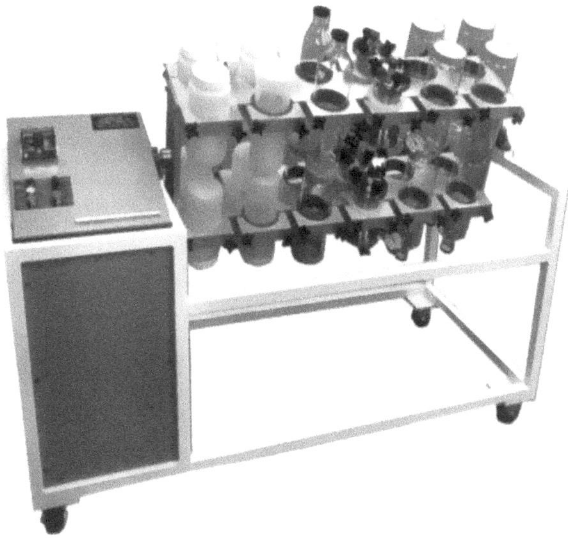

Figure 5.3 Rotation agitator.

- Organics like volatile organics, semivolatile organics, organochlorine insecticides, and polychlorinated biphenyl, organophosphorus pesticides and herbicides.
- Petroleum hydrocarbons, phenols, phthalate esters, etc.

As a proximate analysis, total moisture needs to be determined. The sample that has been exposed to contact with water or wetted by rain may carry free or visible water. This water plus the moisture within the material is referred to as total moisture.

5.3.6 Determination of Total Moisture Content

A known mass of the material is dried and the loss of mass calculated as moisture. The moisture may be determined either by drying in one stage at 108 $\pm 2°C$ or by a two-stage process in which the sample is first air dried under atmospheric conditions and the remaining moisture removed by drying in an oven at 108 $\pm 2°C$. In the latter case, the total moisture is calculated from the loss during air drying and that oven drying. For the determination of moisture in the sample the free-space oven method with temperature of 200 $\pm 5°C$ and a heating period of 4 hours is employed. Total moisture content of the original sample $= X + Y(1 - X/100)$ where

$X = \%$ loss in mass of the original sample in air drying

$Y = \%$ loss in mass of the air dried sample on oven drying

Determination of Volatile Matter: The method consists of heating out of contact with air, a weighed quantity of air-dried sample of hazardous waste at the temperature of 900 $\pm 10°C$ for a period of 7 minutes. Oxidation has to be avoided to the weighed material to secure a nonoxidizing atmosphere.

Volatile matter $= [100(M2 - M3)/M2 - M1] - M0$

where $M0 = \%$ of moisture in sample on air dried basis

$M1 =$ mass in g of empty crucible and lid

$M2 =$ mass in g of empty crucible and lid and sample before heating

$M3 =$ mass in g of empty crucible and lid and sample after heating

5.3.7 Analysis of Hazardous Waste

There are two types of analyses to be carried out for the hazardous waste at two instances, namely, comprehensive analysis by generator at industry and fingerprint analysis by operator of common hazardous waste treatment, storage, and disposal facility.

5.3.7.1 Comprehensive Analysis

It is the responsibility of the generator of waste to do a comprehensive analysis for each kind of waste. This will give an idea of whether the waste is hazardous or not, and also infers the disposal pathway of the waste. The list of parameters to be analyzed is as follows:

- Physical parameters such as physical state (liquid/slurry/sludge/semi-solid/solid (description of different phases)), color, and texture, whether the waste is multilayered (Yes/No; if yes, quantify each layer, specific gravity, viscosity, calorific value, flash point, loss on ignition at 105°C and at 550°C (loss on ignition and paint-filter liquid test).

- Chemical parameters (inorganic): pH, reactive cyanide, reactive sulfide, sulfur (elemental), individual inorganic constituents as relevant (as per Schedule 2 of Hazardous Waste (Management and Handling) Rules, 1989, as amended).
- Chemical parameters (organic): oil and grease, extractable organics (in special cases only), carbon, nitrogen, sulfur, and hydrogen (in percentages), individual organic constituents as relevant (as per Schedule 2 of Hazardous Waste (Management & Handling) Rules, 1989, as amended).
- TCLP for mercury, cadmium, chromium, arsenic, silver, barium, selenium, lead, 2,4-dinitro toluene, vinyl chloride, methyl ethyl ketone, hexachlorobutadiene, 1,2-dichloro ethane, trichloro ethylene, chloro benzene, benzene, hexachloro ethane, carbon tetra chloride, pyridine, chloroform, 1,1-dichloro ethylene, tetrachloro ethylene, nitro benzene, 1,4-dichlo benzene, 2,4,6-trichloro phenol, pentachloro phenol, cresol, *o*-cresol, *m*-cresol, *p*-cresol, 2,4,5-trichloro phenol, heptachlor (and its epoxide), endrine chlordane, hexachloro benzene, lindane, and toxaphene.

5.3.7.2 Fingerprint Analysis

After receiving the waste at the treatment storage and disposal facility (TSDF), the first thing is to check the manifest, Transport Emergency (TREM) card and the other relevant documents. Then, the waste is weighed to confirm the quantities mentioned in the manifest. After that, the truck is moved to the sampling bay to collect samples; depending on the type and condition of the waste, as many numbers of samples have to be collected to get a reasonably representative sample. If the waste is multilayered and seems to be mixed with different varieties, all different samples have to be collected and analyzed separately. The basic objective of the fingerprint analysis is to understand whether the same waste as per the comprehensive analysis report furnished along with the waste has been received or not. Selected parameters (list given later) will normally be carried out quickly because the truck/container has to be kept pending/waiting till the disposal pathway is given by the laboratory. The methodology adopted to quickly conclude on the disposal pathway based on the waste acceptance criteria is physical verification with reference to the master sample, followed by compatibility parameters (spot test methods), detailed analysis (parameters as per the requirement) and comparison with the comprehensive analysis report.

For fingerprint analysis the following tests are performed:

- Physical (physical state—liquid/slurry/sludge/semisolid/solid (description of different phases)), color, and texture, whether the waste is multilayered - (yes/no); if yes, quantify each layer, specific gravity, viscosity, calorific value, flash point, paint-filter liquid test, and loss on ignition at 105°C and at 550°C.

- Chemical parameters such as pH, reactive cyanide, reactive sulfide, chemical compatibility tests or any other test as per necessity based upon the physical, chemical, reactive, toxic, and explosive nature, the wastes have to be segregated at source. Segregation at the source is one of the important elements of the waste management, as it avoids the reaction between different chemical compositions of the products. Segregation based upon its status, whether it is liquid, solid or semi-solid, has to be made. The dry wastes should be segregated from the wet wastes to avoid chemical reactions or accidents. The identification of characteristics of individual waste components is important to segregate the wastes based upon the size of the waste material, moisture content, and density of the liquid or sludge. Schedules to the Rules provide guidelines for segregation of the waste into 18 categories. If the waste generated falls into more than one category, it should be segregated into one of them based upon its reactivity or toxicity. The segregation of the waste should also be made based upon the usage of the waste, i.e., the segregation may be made into reusable waste, recycle waste, and waste without further use. The reusable and recycle wastes have commercial value and can be sold according to the provisions of the law. All the material generated should be segregated immediately on regular intervals without negligence to avoid the mixing of wastes of various characteristics and avoid the accidents. The occupier is responsible for handling of the waste. He/she should take enough care to avoid physical injuries during the segregation. The staff concern should be trained and provided with the equipment to safeguard themselves from the accidents. At the time of segregation, the data of the waste generated at different points should be noted along with its characteristics so that it will be useful for lifting and collection at different stages. Once the segregation is made the waste should be collected by using the proper tins, boxes, or bottles. All the tins, boxes, and bottles should be properly labeled with the detailed composition of the chemicals that are collected without giving the formula alone. The petroleum products and other used oils or inflammable wastes should be collected into tins with caution symbols and placed in the safe places away from fire, and notices should be placed to that extent. The wastes of different categories collected in different tins and different timings and to be placed at different sites. The material used for collection should not be torn or broken, which would facilitate the escape of waste or spreading of the contents from one place to other. The wastes segregated should be collected in regular interval and placed at some specific places that all the workers do not have access to. This is the transitory place for further transporting the wastes to waste-disposal sites. The contents used to collect the wastes should have the information about the type of waste, date of generation, quantity, and quality of the waste collected. The collection of the waste should be made mechanically and manual collection should be avoided. Proper equipment for collection should be used for collection and should not be allowed to be accessed by all other staff of the industry. A trained supervisor should be appointed to supervise the segregation and collection activities of the industry.

5.3.8 Equipment/Instruments for Hazardous Waste Analysis and Characterization

Equipment and instruments for hazardous waste analysis and characterization include moisture/water content analyzer, bomb calorimeter, flash point apparatus, TCLP agitator, microwave digestion system, Atomic Absorption Spectrophotometer with Hydride Vapor Generation/ Inductively Coupled Plasma Atomic Emission Spectrophotometer/ Inductively Coupled Plasma Mass Spectrophotometer (AAS-HVG/ICP-AES/ICP-MS), Zero Headspace Extractions (ZHE), gas chromatograph, Gas Chromatography Mass Spectrophotometer (GC-MS) and Carbon-Hydrogen-Nitrogen- Sulphur (CHNS) analyzer, analytical balance, bulk density apparatus/pycnometer, viscometer, hot-air oven, KF titrator, muffle furnace, paint filter papers, pH meter, spectrophotometer, High Performance Liquid Chromatography (HPLC) (for specific parameters), Adsorbable Organic Halides (AOX) analyzer, etc.

5.4 STORAGE, TRANSPORT, AND DISPOSAL OF HAZARDOUS WASTE

After identifying and characterizing the waste as hazardous waste, it needs special storage, transport, and disposal. The details are as follows:

5.4.1 Storage of Hazardous Waste

The *storage of hazardous waste* means the keeping of hazardous waste for a temporary period, at the end of which the hazardous waste is treated and disposed of. The hazardous waste should not be stored at the place of the generation. It may be allowed provided it is stored in the specific containers used for storage, the date of storage is stated specifically, and each container contains the label and tag of the waste stored. On the day of generation of the waste, it should be segregated and collected from different departments of the industry and sent for the transit storage for facilitating the storage at the place of "earmarked" storage point. The storage area should be away from the place of generation and specifically marked for the purpose of storage. The storage areas should be cleaned at fixed intervals. The containers with leakage should be avoided. Space should be made available in between the containers of different characteristics. The storage area should have all the equipment for controlling the pollution, water-spray systems, and alarm systems to caution others. Regular inspections have to be carried out to find the deficiencies of the storage systems. An extra number of containers should be available at the place of storage to meet the exigencies of the demand or excess generation of the wastes. Some of the important tips for the storage of hazardous waste are given below:

- Hazardous waste should be stored in sealed, compatible containers.
- Hazardous waste containers must be kept closed at all times except to add waste.
- Label hazardous waste containers with tags as soon as waste accumulation begins.
- Store hazardous wastes with secondary containment.
- Segregate incompatible hazardous wastes into different containers.
- Never accumulate more than 60% of the capacity of the container.
- Ensure that lab personnel are trained on proper waste handling procedures.
- Maximum quantity permissibility depends upon the type and character of the waste intended to be stored.
- The maximum quantity should not be more than a truckload.
- The maximum period of storage should be for a period of 90 days or the quantity collected, whichever is earlier, and under special circumstances only may storage beyond 90 days be permitted.
- Hazardous chemicals should be stored very cautiously as they are subject to easy chemical reactions and explosions. The content and composition of the chemicals should be clearly mentioned instead of adding the formulae.
- Indefinite and uncontrolled storage of chemicals may cause problems.
- Use of branded containers or containers meeting the specific standards is preferred for storage.

There should be a temporary storage place with in the TSDF to retain the waste in case there is a delay in deciding the disposal pathway by the laboratory. Temporary storage is also necessary during the rainy period to temporarily store the waste under covered and impervious line sheds. There are also occasions that the waste characteristics may permit neither landfilling nor incineration. In these circumstances and in the absence of proven and accepted disposal methodologies, the only option left for these wastes is a long-term storage. These types of wastes are named as intractable wastes, hence, an intractable waste storage has to be provided in the facility.

5.4.2 Transport of Hazardous Waste

To transport hazardous waste, the industry may have its own transport system or have arrangements with common treatment, storage, and disposal facilities. Transport means movement of hazardous waste by air, rail, road, or water and transporter means a person engaged in the offsite transportation of hazardous waste. The operator or the transporter should have license or authorization from the Pollution Control Board/Pollution Control Committees or registered with Ministry of Environment and Forests, Government of India. The drivers/cleaners/helpers should be trained in handling the waste in case of exigencies or emergency situations. The transporting should be undertaken by specified dedicated vehicles only. The vehicle used for transportation should be leak-proof and have the capacity to take away the load. Spill risk is high during loading, transportation, and unloading. The vehicle should have a notice for use of waste disposal and should have the notice of address of the transporter along with phone number. It should contain first-aid required in case of accidental spills. It should carry licenses and permit and No Objection Certificate documents. The packages and containers should be handled with care to avoid breakage during transportation or loading and unloading. Containers used during the transport shall be of mild steel with suitable corrosion-resistant coating and roll-on roll-off cover, which may either be handled by articulated crane or by a hook-lift system comfortably for a large variety of wastes. Other modes of packaging, like collection in 200-L plastic drums, cardboard cartons, PP and HDPE/LDPE containers, etc., also work for variety of wastes. However, all such containers should be amenable to mechanical handling. In general, the containers for liquid hazardous waste should be completely closed (in fact sealed). There should be no gas generation due to any chemical reaction within the container, and, hence, there should not be any need for air vents, as expansion due to increase/decrease in temperature normally does not need air vents. The container should be covered with a solid lid or a canvas to avoid emissions of any sort including spillage, dust, etc., and to minimize odor generation both at the point of loading as well as during transportation. The container used for transportation of waste should be able to withstand the shock loads due to vibration effect/undulations of pavement, etc., and should be easy to handle during transportation and emptying. As far as possible, manual handling of containers should be minimized. Appropriate material handling equipment is to be used to load, transport, and unload containers. This equipment includes drums, dollies, and forklifts; drum -handling equipment; lift gates; and pallets. Drums should not be rolled on or off vehicles. Loads are to be properly placed on vehicles. Hazardous waste containers are not to overhang, perch, lean, or be placed on any other unstable base. Load should be secured with straps, clamps, braces, or other measures to prevent movement and loss. Design of the container should be such that it can be safely accommodated on the transport vehicle. Dissimilar wastes shall not be collected in the same container. Wastes shall be segregated and packed separately. This is necessary to ensure that each waste finds its way to the right disposal point. All hazardous waste containers must be clearly marked with current contents. The markings must be waterproof and firmly attached so that they cannot be removed. Previous content labels shall be obliterated when the contents are different. Proper marking of containers is essential. It should be labeled with the words HAZARDOUS WASTE in the vernacular language, Hindi, and English. The information on the label must include the code number of the waste, the waste type, the origin (name, address, telephone number of generator), hazardous property (e.g., flammable), and the symbol for the hazardous property (e.g., the red square with flame symbol). The label must withstand the effects of rain and sun. Labeling of containers is important for tracking the wastes from the point of generation up to the final point of disposal. The following are the requirements for labeling:

- The label should contain the name and address of the occupier and operator of the facility where it is being sent for treatment and final disposal, i.e., labeling of container shall be provided with a general label as per Form 8 of the HW (M & H) Rules, 1989, and as amended.
- Emergency contact phone numbers shall be prominently displayed viz. the phone number of concerned regional officer of the SPCB/PCC, fire station, police station, and other agencies concerned.
- All hazardous waste containers shall be provided with a general label as given in Form 8 in Hazardous Waste (Management & Handling) Rules, 1989, as amended.

- Transporter shall not accept hazardous wastes from an occupier (generator) unless six copies (with color codes) of the manifest (Form 9) as per Rule 7 of the Hazardous Waste (Management & Handling) Rules, 1989, and as amended, is provided by the generator. The transporter shall give a copy of the manifest signed and dated to the generator and retain the remaining four copies to be used for further necessary action prescribed in the Hazardous Wastes (Management & Handling) Rules, 1989, as under:
  - Copy 1 (white) to be forwarded to the SPCB/PCC by the occupier
  - Copy 2 (yellow) to be signed by the transporter and retained by the occupier
  - Copy 3 (pink) to be retained by the operator of a facility
  - Copy 4 (orange) to be returned to the transporter by the operator of facility after accepting waste
  - Copy 5 (green) to be forwarded to the SPCB/PCC by the operator of facility after disposal
  - Copy 6 (blue) to be returned to the occupier by the operator of the facility after disposal.
- In case of interstate transportation of waste, the occupier (waste generator) shall strictly follow the manifest system as stipulated under Rule 7(5) of the Rules.
- In case of transport of hazardous wastes to a facility for treatment, storage, and disposal existing in a State other than the State where wastes are generated, the generator shall obtain necessary No Objection Certificate from the concerned State Pollution Control Board or Pollution Control Committee of the UT where the facility is located (as stipulated under Rule 7 (6) of Rules.
- Wastes in transport must be in closed containers at all times, and must be delivered to designated points only.
- Vehicles shall be painted preferably in blue color with white strip of 15–30 cm width running centrally all over the body. This is to facilitate easy identification.
- Vehicles should be fitted with mechanical handling equipment as may be required for safe handling and transportation of the wastes.
- The words HAZARDOUS WASTE shall be displayed on all sides of the vehicle in the vernacular language, Hindi, and English.
- Name of the facility operator or the transporter, as the case may be, shall be displayed.
- Carrying of passengers is strictly prohibited and only those associated with the waste haulers shall be permitted in the cabin.

- The trucks shall be dedicated for transportation of hazardous wastes and they shall not be used for any other purpose.
- Educational qualifications for the driver shall be minimum of 10th pass (SSC). The driver of the transport vehicle shall have a valid driving license for heavy vehicles from the State Road Transport Authority and shall have experience in transporting chemicals.

5.4.3 Technologies for Hazardous Waste Management

Various technologies for hazardous waste management have evolved; they are biological treatment, physical treatment, chemical treatment, landfill, and incineration.

5.4.3.1 Biological Treatment

Biological treatment systems use microorganisms for degradation of toxic and hazardous waste. The emerging techniques are efficient and economical for detoxifying contaminant water, soils, and sediments. Naturally occurring microbes and genetically engineered varieties have received much attention for treatment of chlorinated organics. Recent interest has focused on unique hydrogen peroxidase-dependent oxidase secreted by the white rot fungus, *Phanerochaete crysosporium*. It is capable of degrading lignin complex and toxic organohalide pollutants such as 2,4,6-trichlorophenol. The United States has developed bacterial isolates to degrade polychlorinated biphenyls (PCBs). *Alkaligenes eutrophous and Pseudomonas putida* have degraded over 90% of the PCBs in soils. A microbial process for removing pentachlorophenol (PCP) and related polyaromatic hydrocarbon (PAHs) has been developed in the United States. The contaminated soil is slurried with a caustic and water, placed in a digester, and heated to 180°F and agitated for 24 hours. It is cooled and inoculated with *Arthobacter* and stirred for 48 hours and dewatered. The pollutants reduced from 500 to 1 ppm. Various microbial species used for detoxification of selected hazardous xenobiotics are presented in Table 5.1.

5.4.3.2 Physical Treatment

Physical treatment methods are used for removal rather than destruction; these methods use physical characteristics to separate constituents in a waste stream, wherein residues are separated and then further treatment and ultimately disposed off. Gravity separation (sedimentation, centrifugation, flocculation, oil/water separation, dissolved air flotation, and heavy media separation), phase change (evaporation, air stripping, steam

Table 5.1 Microbial species for detoxification of selected hazardous xenobiotics

S.No.	Hazardous pollutants	Microbial species
1.	Phenols	*Achromobacter, Alcaligens, Acinatobacter, Arthobacter, Azotobacter, Bacillus cereus, Flavobacterium, Pseudomonas putida, P. aeruginosa, Nocardia, Candida tropicalis, Trichosporon cutaneum, Aspergillus, Penicillium,* and *Neurospora*
2.	Dyes and dye intermediates	*Bacillus sp., Flarabaderium sp., Pseudomonas sp.*
3.	Hydrocarbon	*Escherichia coli, Pseudomonas putida, P. aeruginosa,* and *Candida*
4.	Pesticides: DDT	*Pseudomonas aeruginosa*
5.	Pesticides: Linuron	*B. sphaericus*
6.	Pesticides: 2,4-D	*Arthrobacter* and *P. cepacia*
7.	Pesticides: 2,4,5-T	*Pseudomonas sp., E. coli*
8.	Parathion	*P. stutuzer* and *P. aeruginosa*
9.	Cyanide	*Bacillus megatherium, B. subtilis, Pseudomonas sp., Arthrobacter sp., Nocardia, Fusarium solani, Aspergillus niger, Rhizopus nigricans,* and *Rhizoctonia solani*
10.	Dioxins	Mutant strain of *Pseudomonas* (sp. NCIB 9816 strain II)

Source: (1) Brainstorming Session, Environmental Biotechnology, Vol. 1, NEERI, April 1989, (2) Kulla, Hg. Aerobic bacterial degradation of azodyes. Microbial degradation of xenobiotics and recalcitrant compounds (FEMS Symp. Ser.12) Academic Press, London, 1981, (3) Dioxins. USEPA-600/2-80-197. Cincinnati, OH.

stripping, and distillation), dissolution (soil washing/flushing, chelation, liquid/liquid extraction, and supercritical solvent extraction), and size/adsorptivity/ionic characteristics (filtration, carbon adsorption, reverse osmosis, ion exchange, and electrodialysis) are various methods employed to treat the waste by physical treatment. To adopt physical treatment, certain data is needed for solids, liquids, solid and liquid mixtures, and gases. The type of data needed is as given in Table 5.2.

5.4.3.3 Physical Treatment Process

5.4.3.3.1 Gravity Separation: Sedimentation Process

Sedimentation is a settling process in which gravity causes heavier solids to collect at the bottom of containment vessel, separated from the suspending fluid. The sedimentation process can be batch process or continuous removal process. The important sedimentation data needed is as given in Table 5.3.

Table 5.2 Important physical treatment data needs for adopting the physical treatment process

S.No.	Data need	Purpose
For solids		
1.	Absolute density	Density separation
2.	Bulk density	Storage volume required
3.	Size distribution	Size modification or separation
4.	Friability	Size reduction
5.	Solubility (in H_2O, organic solvents, oils, etc.)	Dissolution
For liquids		
1.	Specific gravity	Density separation
2.	Viscosity	Pumping and handling
3.	Water content (or oil content, etc.)	Separation
4.	Dissolved solids	Separation
5.	Boiling point/freezing point	Phase-change separation, Handling and storage
For liquid/solid mixtures		
1.	Bulk density	Storage and transportation
2.	Total solids content	Separation
3.	Solids size distribution	Separation
4.	Suspended solids content	Separation
5.	Suspended solids settling rate	Separation
6.	Dissolved solids content	Separation
7.	Free water content	Storage and transport
8.	Oil and grease content	Separation
9.	Viscosity	Pumping and handling
For gases		
1.	Density	Separation
2.	Boiling (condensing) temperature	Phase-change separation
3.	Solubility (in H_2O, etc.)	Dissolution

Table 5.3 Important sedimentation data

Data need	Purpose
Viscosity of aqueous waste	High-viscosity hinders sedimentation
Oil and grease content of waste stream	Not applicable to wastes containing emulsified oils
Specific gravity of suspended solids	Must by greater than 1 for sedimentation to occur

The settling pond is shown in Fig. 5.4; circular clarifier and basin sedimentation are various types of sedimentation processes. In the settling pond, aqueous waste flows through while suspended solids are permitted to gravitate and settle out and occasionally the settling particles (sludge) are removed.

The circular clarifier (Fig. 5.5) is equipped with a solids-removal device. The solids-removal device facilitates continuous clarification, resulting in a lower solid-content outlet fluid.

The sedimentation basin (Fig. 5.6) uses a belt-type solids collector mechanism to force solids to the bottom of the basin's sloped edge, where the solids are moved. The efficiency of sedimentation treatment depends upon the depth and surface area of the basin, settling time (based on the holding time), solid particle size, and flow rate of the fluid.

Sedimentation is considered a separation process only. Typically, some type of treatment process for aqueous liquids and sludges will follow. Its limitations are as follows: use is restricted to solids that are more dense then water, and it is not suitable for wastes consisting of emulsified oils.

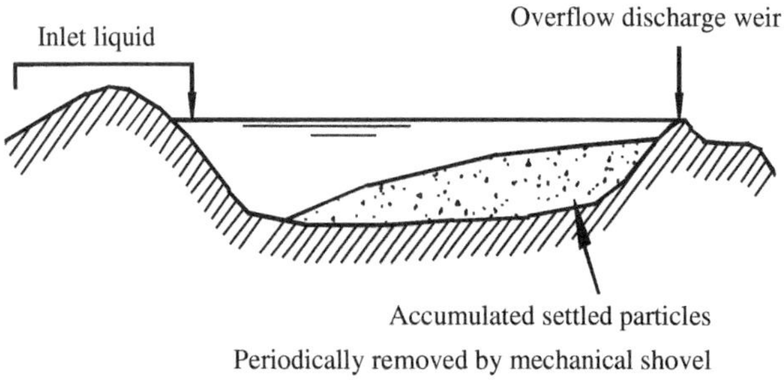

Figure 5.4 Settling pond.

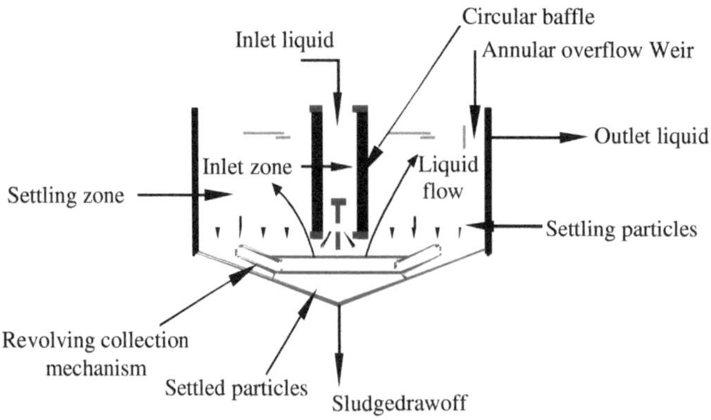

Figure 5.5 Circular clarifier.

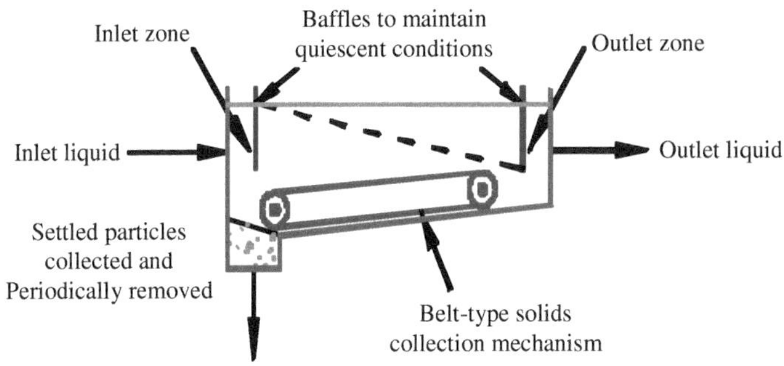

Figure 5.6 Sedimentation basin.

5.4.3.4 Gravity Separation: Centrifugation

Centrifugation involves physical separation of fluid-mixture components based on their relative density.

A rapidly rotating fluid mixture within a rigid vessel deposits denser solid particles farther from the axis of rotation, while liquid supernatant lies separated near the axis.

Centripetal forces in centrifugation are similar to gravitational forces in sedimentation, except the centripetal forces are thousands of times stronger than gravitational forces, depending upon centrifuge diameter and rotational speed. Centrifugation is limited to dewatering of sludge (including metal-bearing sludge), separating oils from water, and clarification of viscous gums and resins. Centrifuges are generally better suited than vacuum filters for dewatering sticky or gelatinous sludge. Disk-type centrifuges can be used to separate three component mixtures (e.g., oil, water, and solids). The limitations for this process include the fact that centrifuges often cannot be used for clarification since they may fail to remove less-dense solids and those small enough to remain in suspension. Recovery and removal efficiencies can be improved if paper or cloth filters are used (Fig. 5.7).

5.4.3.5 Separation: Flocculation

Flocculation is used primarily for the precipitation of inorganic substances. It is used to enhance sedimentation or centrifugation. The waste stream is mixed while a flocculating chemical is added. Flocculants adhere readily to suspended solids and to each other (agglomeration), and the resultant particles are too large to remain in suspension. Flocculation is a conventional, demonstrated treatment technique. The extent of flocculation depends upon waste stream flow rate, composition, and pH. Limitation being this process is not recommended for a highly viscous waste stream (Table 5.4).

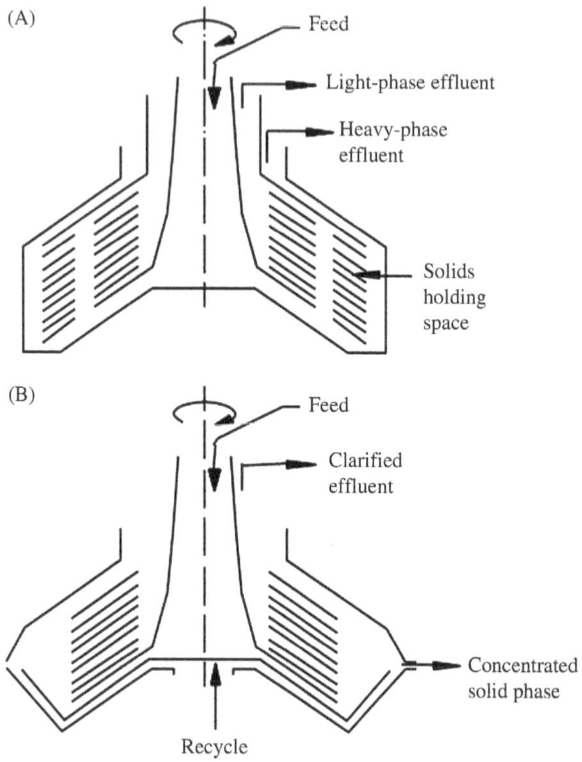

Figure 5.7 Disk-centrifuge bowls. (A) Separator, solid wall. (B) Recycle clarifier, nozzle discharge.

Table 5.4 Important flocculation data needs

Data need	Purpose
pH of waste	Selection of flocculant agent
Viscosity of waste system	Affects settling of agglomerated solids; high viscosity is not suitable
Settling rate of suspended solids	Selection of flocculating agent

5.4.4 Gravity Separation: Oil/Water Separation

The force of gravity can be used to separate two or more immiscible liquids with sufficiently different densities, e.g., oil and water. Liquid/liquid separation occurs when the liquid mix settles. Flow rates in continuous processes must be kept low. The waste flows into a chamber, where it is kept quiescent and permitted to settle. The floating oil is skimmed off the top using an oil skimmer while the water or effluent flows out of the lower portion of the chamber. Acids may be used to break oil/water emulsion and to enhance this process for efficient oil removal. Effectiveness can be influenced by waste-stream flow rate, temperature, and pH. Separation is a pretreatment process if the skimmed oil requires further treatment. Mobile phase separators are commercially available.

5.4.5 Gravity Separation: Dissolved Air Flotation (Fig. 5.8)

Dissolved air flotation involves removing suspended particles or mixed liquids from an aqueous waste stream. The mixture to be separated is saturated with air or another gas such as nitrogen, and then air pressure is reduced above the treatment tank. As air escapes the solution, microbubbles form and are readily adsorbed onto suspended solids or oils, enhancing their flotation characteristics. In the flotation chamber, separate oil or other floats are skimmed off the top while aqueous liquids flow off the bottom. It is only applicable for waste with densities close to water. Air-emission controls may be necessary if hazardous volatile organics are present. This is a conventional treatment process.

5.4.5.1 Gravity Separation: Heavy Media Separation

Heavy media separation is used to process two solid materials with significantly different absolute densities. Mixed solids are placed in a fluid with a

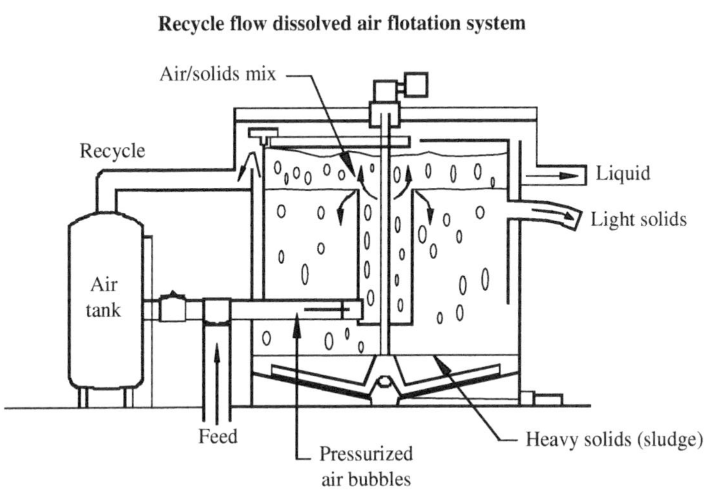

Figure 5.8 Gravity separation: dissolved air flotation.

specific gravity adjusted to allow lighter solids to float while heavier solids sink. Usually, the separating fluid or heavy medium is a suspension of magnetite in water. The specific gravity is adjusted by varying the amount of magnetite powder used. Magnetite is easily recovered magnetically from rinse waters and spills, then reused. This type of separation is used to separate two insoluble solids with different densities. Limitations include the possibility of dissolving solids and ruining the heavy media; the presence of solids with densities similar to those solids requiring separation; and the inability to cost-effectively separate magnetic materials, because of the need to recover magnetite. Commonly used in the mining-industry to separate ores from tailings.

5.4.5.2 Phase Change: Evaporation (Fig. 5.9)

Evaporation is the physical separation of a liquid from a dissolved or suspended solid by applying energy to make the liquid volatile. Evaporation may be used to isolate the hazardous material in one of the two phases, simplifying subsequent treatment if the hazardous waste is volatilized; this

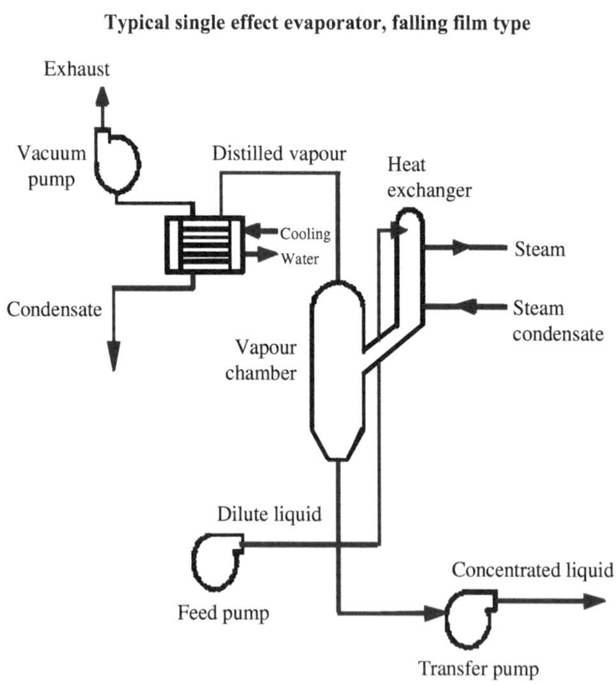

Figure 5.9 Phase change: evaporation.

process is usually called stripping. Evaporation can be applied to any mixture of liquids and volatile solids provided the liquid is volatile enough to evaporate. If the liquid is water, evaporation can be carried out in large ponds using solar energy. Aqueous waste can also be evaporated in closed-process vessels using steam energy. The resulting water vapor can be condensed for reuse. Energy requirements are minimized by techniques such as vapor recompression or multiple-effect evaporators. Evaporation is applied to solvent waste contaminated with nonvolatile impurities such as oil, grease, paint solids, or polymeric resins. Solvent is evaporated and recovered for reuse. This process is commercially available.

5.4.5.3 Phase Change: Air Stripping

Air stripping is a mass transfer process in which volatile contaminants in water or solids are evaporated into the air. Organics removal by air stripping depends upon temperature, pressure, air-to-water ratio, and surface area available for mass transfer. Volatile hazardous material must be recaptured for subsequent treatment to preclude air pollution. This process is used to treat aqueous wastes that are more volatile, less soluble (e.g., chlorinated hydrocarbons such as tetrachloroethylene), and aromatic (e.g., toluene). The limitations are that temperature dependency and presence of suspended solids may reduce efficiency. If the concentrations of volatile organic contaminants (VOCs) exceeds about 100 ppm, steam stripping is preferred. This process is commercially available.

5.4.6 Phase Change: Steam Stripping

Steam stripping (Fig. 5.10) uses steam to evaporate volatile organics from aqueous wastes. It is essentially a continuous fractional distillation process carried out in a packed or tray tower. Clean steam, rather than reboiled bottoms, provide direct heat to the column, and gas flows from the bottom to the top of the tower. The resulting residuals are contaminated steam condensate, recovered solvent, and stripped effluent. The organic vapors are sent through a condenser for further purification treatment. The bottom requires further consideration as well. Possible posttreatment includes incineration, carbon adsorption, or land disposal.

Steam stripping is used to treat aqueous wastes contaminated with chlorinated hydrocarbons, aromatics such as xylenes, ketones such as acetone, alcohols such as methanol, and high-boiling point chlorinated

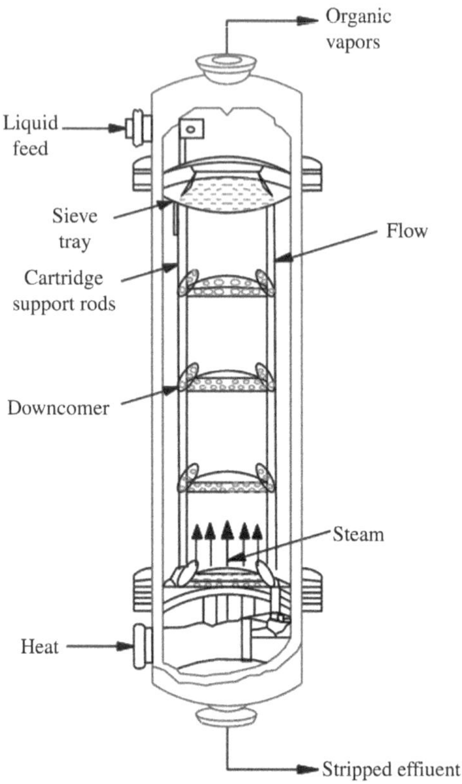

Figure 5.10 Steam-stripping column (perforated tray type).

aromatics such as pentachlorophenol. Steam stripping will treat less volatile and more soluble wastes, can handle a wide concentration range (e.g., from less than 100 ppm to about 10% organics), and requires an air-pollution control device to eliminate toxic emissions. It is a conventional method and is well documented.

5.4.6.1 Phase Change: Distillation

Distillation is simply evaporation followed by condensation. The separation of volatile materials is optimized by controlling the evaporation-stage temperature and pressure and the condenser temperature. Distillation separates miscible organic liquids for solvent reclamation and waste-volume reduction. The two types of distillation processes are batch distillation and continuous fractional distillation. Distillation is used to separate liquid organic wastes, primarily spent solvents, for full or partial recovery and reuse. Liquids to be separated must have different volatilities. Distillation for recovery is limited by the presence of volatile or thermally reactive suspended solids. Batch distillation in a heated still pot with condensation of overhead vapors is easily controlled and flexible, but cannot achieve the high product quality typical of continuous fractional distillation. Continuous distillation is accomplished in tray columns or packed columns ranging up to 40 ft in diameter and 200 ft high, each is equipped with a reboiler, a condenser, and an accumulator. Fractional distillation is not applicable to liquids with high viscosity at high temperature, liquids with high solid concentrations, polyurethanes, or inorganic. It is commercially available.

5.5 DISSOLUTION: SOIL FLUSHING/SOIL WASHING (FIG. 5.11)

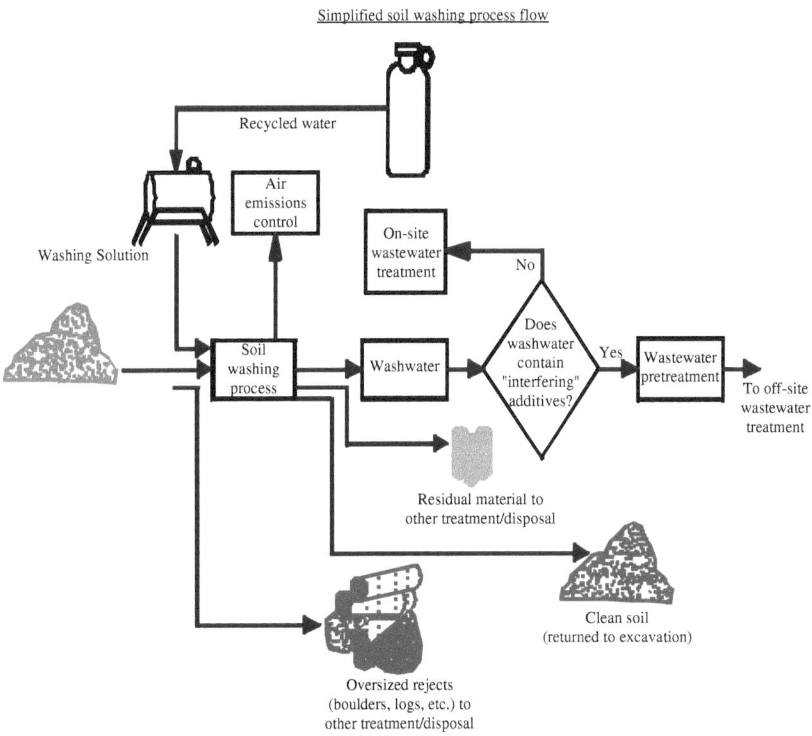

Figure 5.11 Dissolution: soil flushing/soil washing.

5.5.1 Soil Flushing

Soil flushing is an in situ extraction of inorganic and organic compounds from soil, and is accomplished by passing extractant solvents through the soils using an injection and recirculation process. Solvents includes water, water-surfactant mixtures, acids, bases (for inorganics), chelating agents, and oxidizing agents or reducing agents.

5.5.2 Soil Washing

Soil washing consists of excavating and treating at the surface in a soil washer. The contaminated soil is removed to a staging area, then shifted to remove debris and large objects such as rocks; the remaining material enters a soil scrubbing unit is mixed with a washing solution and agitated. The washing solution may be water or may contain some additives like detergent to remove contaminants. Then, the wash water is drained and the soil is rinsed with clean water. The wash water, which also contains contaminants, undergoes wastewater treatment processes for future recycling use. Soil flushing and washing fluids must have good extraction coefficients, low volatility and toxicity, capability for safe and easy handling, and most important be recoverable and recyclable. This technology is very promising in extracting heavy metals from soil, although problems are likely in dry or organically rich soils. Surfactants can be used to extract hydrophobic organisms. Soil type and uniformity are important. Certain surfactants, when rested for in situ extraction, clogged soil pores and precluded further flushing.

5.5.3 Dissolution: Chelation

A chelating molecule contains atoms that form ligands with metal ions. If the number of such atoms in the molecule is sufficient, and if the molecular shape is such that the final atom is essentially surrounded, then the metal will be unable to form ionic salts that can precipitate out. Chelation is used to keep metals in solution and to aid in dissolution for subsequent transport and removal (e.g., soil washing). Chelating chemicals are chosen for their affinity to particular metals and the presence of fats and oils can interfere with the process. Chelating chemicals are commercially available.

5.5.4 Dissolution: Liquid/Liquid Extraction

Two liquids that are well mixed or mutually soluble may be separated by liquid/liquid extraction. The process requires that a third liquid be added to the original mix. This third liquid must be a solvent for one of the original components, but must be insoluble in and immiscible with the other. The final solvent and solute stream can be separated by distillation or other chemical means, and the extracting solvent captured and reused. Complete separation is rarely achieved, and some form of posttreatment is required for each separated stream. To effectively recover solvent and solute materials from the process, other treatments such as distillation or stripping are needed. This is a demonstrated process.

5.5.5 Dissolution: Supercritical Extraction

At a certain temperature and pressure, fluids reach their critical point, beyond which their solvent properties are greatly enhanced. For instance, supercritical water is an excellent nonpolar solvent in which most organics are readily soluble. These properties make extraction more rapid and efficient than distillation or conventional solvent extraction methods. Presently, the use of supercritical carbon dioxide to extract hazardous organics is being investigated. This technology may be useful in extracting hazardous waste from aqueous streams. Specific applicability and limitations are not yet known. This process has been demonstrated on a laboratory scale.

5.5.6 Size/Adsorptivity/Ionic Characteristics: Filtration

Filtration is the separation and removal of suspended solids from a liquid by passing the liquid through a porous medium. The porous medium may be a fibrous fabric (paper or cloth), a screen, or a bed of granular material. The filter medium may be precoated with a filtration aid such as ground cellulose or diatomaceous earth. Fluid flow through the filter medium may be accomplished by gravity, by including a partial vacuum on one side of the medium, or by exerting mechanical pressure on a dewaterable sludge enclosed by filter medium. Filtration is used to dewater sludges and slurries as pretreatment for other processes. It is also a polishing step for treated waste, reducing suspended solids and associated contaminants to low levels. Pretreatment by filtration is appropriate for membrane separation, ion exchange, and carbon adsorption to prevent plugging or overloading the processes. The limitations are that filtration of settled waste is often required to remove undissolved heavy metals present as suspended solids. Filtration does not reduce waste toxicity, although powdered activated carbon may be used as an absorbent and filter aid and it should not be used with sticky or gelatinous sludges, due to the likelihood of filter media plugging. This process is commercially available.

5.5.7 Size/Adsorptivity/Ionic Characteristics: Carbon Adsorption

Most organic and inorganic compounds will readily attach to carbon atoms. The strength of that attachment and the energy for subsequent desorption depends on the bond formed, which in turn depends on the specific compound being adsorbed. Carbon used for adsorption is treated to produce a high surface-to-volume (900: 1,300 sq. m/g), exposing a practical maximum number of carbon atoms for active adsorption. This treated carbon is said to be activated for adsorption. When activated carbon has adsorbed so much contaminant that its adsorptive capacity is severely depleted, it is said to be spent. Spent carbon can be regenerated, but for strongly adsorbed contaminants, the cost of such regeneration is higher than simple replacement with new carbon. This process is used to treat single-phase aqueous organic wastes with high molecular weight and boiling point, and low solubility and polarity, chlorinated hydrocarbons such as tetrachloroethylene, and aromatics such as phenol. It is also used to capture volatile organics in gaseous mixtures. Limitations are economic, relating to how rapidly the carbon becomes spent. As an informal guide, concentrations should be less than 10,000 ppm, suspended solids less than 50 ppm; and dissolved inorganics, oil, and grease less than 10 ppm. It is conventional and demonstrated.

5.5.8 Size/Adsorptivity/Ionic Characteristics: Reverse Osmosis

In normal osmotic processes, solvent flows across a semipermeable membrane from a dilute solution to a more concentrated solution until equilibrium is reached. Applying high pressure to the concentrated side causes the process to reverse. Solvent flows from the concentrated solution, leaving an even higher concentration of solute. The semipermeable membrane can be flat or tubular, and acts like a filter due to the pressure-driving force. The waste stream flows through the membrane, while the solvent is pulled through the membrane's pores. The remaining solutes, such as organic or inorganic components, do not pass through, but become more and more concentrated on the influent side of the membrane. For efficient reverse osmosis, the semipermeable membrane's chemical and physical properties must be compatible with the waste stream's chemical and physical characteristics. The limitations are that some membranes will be dissolved by some wastes. Suspended solids and some organics will clog the membrane material. Low-solubility salts may precipitate onto the membrane surface. Commercial units are available.

5.5.9 Size/Adsorptivity/Ionic Characteristics: Ion Exchange

The ion-exchange process normally uses specially formulated resins with an exchangeable ion bonded to the resin with a weak ionic bond. Ion exchange depends upon the electrochemical potential of the ion to be recovered versus that of the exchange ion and the concentration of the ions in the solution. After a critical relative concentration of recoverable ion to exchanged ion in the solution is exceeded, the exchanged resin is said to be spent. Spent resin is usually recharged by exposure to a concentrated solution of the original exchange ion, causing a reverse exchange. This results in regenerated resin and a concentrated solution of the removed ion, which can be further processed for recovery and reuse. The ion-exchange process is used to remove toxic metal ions from solution to recover concentrated metal for recycling. The residuals include spent resins and spent regenerate such as acid, caustic, or brine. This technology is used to treat metal wastes including cations (e.g., Ni^{2+}, Cd^{2+}, Hg^{2+}) and anions (e.g., $CrO^{2-}{}_4$, $SeO^{-2}{}_4$, $HASO^{2-}{}_4$). The limitations are concentrated waste streams with greater than 25,000 mg/L contaminants can be more cost-effectively separated by other means. Solid concentrations greater than 50 mg/L should be avoided to prevent resin blinding. This is a commercially available process.

5.5.10 Size/Adsorptivity/Ionic Characteristics: Electrodialysis

In electrodialysis (Fig. 5.12), a water solution is passed through alternately placed cation-permeable and anion-permeable membranes. An electrical potential is applied across the membrane to provide the motive force for ion migration. The ion-selective membranes are thin sheets of ion-exchange resins reinforced by a synthetic fiber backing. Applied for purifying brackish water, this method was demonstrated for recovery of metal salts from plating rinse. The units are being marketed to reclaim metals of value from rinse streams. Such units can be skid-mounted and require only piping and electrical connections.

5.5.10.1 Chemical Treatment

Chemical treatment processes alter the chemical structure of wastes, producing residuals that are less hazardous than the original waste.

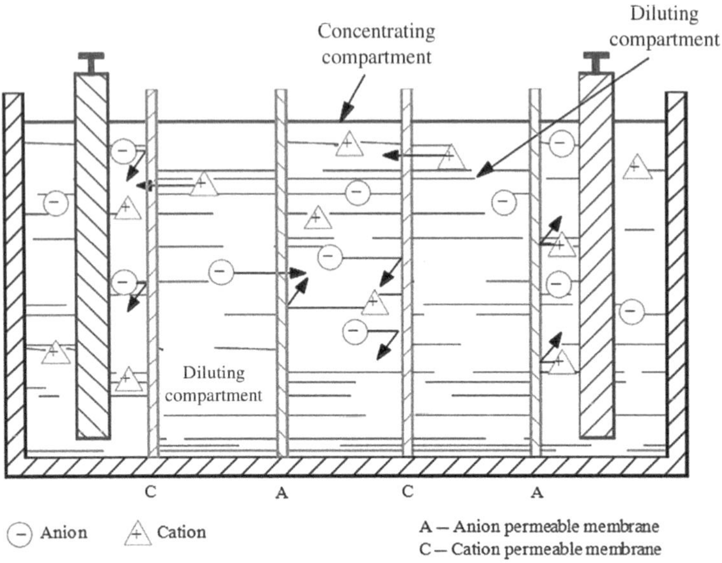

Electrodialysis. An electric current concentrates the dissolved ions in compartments adjacent to those between the electrodes.

Figure 5.12 Electrodialysis.

Various (commonly used) chemical treatment processes are pH adjustment (for neutralization or precipitation), oxidation and reduction, hydrolysis and photolysis, chemical oxidation (ozonation, electrolytic oxidation, hydrogen peroxide), and chemical dehalogenation (alkaline metal dechlorination, alkaline metal/polyethylene glycol, based catalyzed dechlorination).

5.5.10.2 pH Adjustment: Neutralization (Fig. 5.13)

This is an inexpensive treatment. An ionic salt is dissolved in water--→ H_2O --→H^+, OH^-. In neutralization change in the constituents in an ionic solution until ions $\mathbf{OH^- = H^+}$ takes place. Neutralization is used to treat waste acids and alkalis to treat each other and to eliminate/reduce reactivity and corrosivity. The residuals (end products) include neutral effluence-containing dissolved salts and any precipitated salts. pH adjustment has a wide application like aqueous and nonaqueous liquids, slurries, and sludge. The applications include pickle liquors, plating wastes, mine drainage, and oil-emulsion breaking. There will not be any change in physical form, except precipitation or gas evolution. It should be performed in a well-mixed system to ensure completeness. Compatibility of

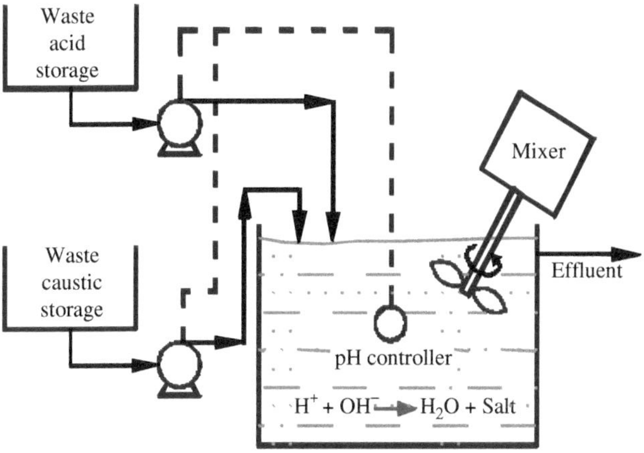

Figure 5.13 Simultaneous neutralization of acid and caustic waste.

the waste and treatment chemicals should be ensured to prevent formation of more toxic or hazardous compounds than were originally present. This is a common industrial process.

5.5.10.3 Chemical Precipitation

Chemical precipitation is a pH adjustment process mainly used for removal of dissolved metals from aqueous wastes. An acid or base is added to a solution to adjust the pH to a point where the constituents to be removed reach their lowest solubility. The solubility of metals decreases as pH increases and the metal ions precipitate out of the solution as hydroxide. Metals can be precipitated by adding alkaline agents, such as lime or caustic soda to raise pH. To remove

Chromium → Sodium bisulfate

Cyanide → Sulfates, including zinc sulfate or ferrous sulfate

Metals → Carbonates, especially calcium carbonate

are used.

Hydroxide precipitation with lime is most common sodium sulfide and is sometimes used to achieve lower effluent metal concentrations. The residuals are metal sludge/treated effluent with an elevated pH/sulfide precipitation, excess sulfide. This method is applied to treat aqueous wastes containing metals. The limitations are complexing agents can interfere with the process.

Organics are not removed except through adsorptive carryover. The resulting sludge may be hazardous. Precipitation has many useful

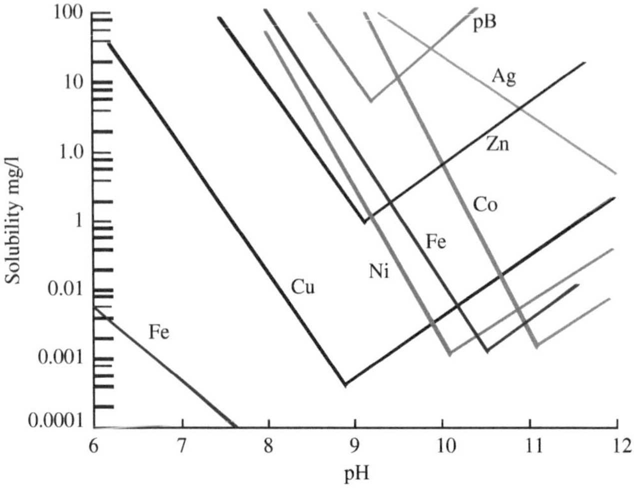

Figure 5.14 Solubilities of metal hydroxide at various pH.

applications to hazardous waste treatment, but laboratory jar tests should be made to verify the treatment. The jar test is used to select the appropriate chemical; determine dosage rates; assess mixing, flocculation, and settling characteristics; and estimate sludge production and handling requirements (Fig. 5.14).

5.5.10.4 Oxidation and Reduction

Oxidation and reduction change the chemical form of a hazardous material, end products are less toxic, changing its solubility, stability, or separability, or otherwise changing it for handling or disposal purposes. In any oxidation reaction, the oxidation state of one compound is raised while the oxidation state of another compound is reduced. The compound supplying oxygen, chlorine, or another negative ion, is called the oxidizing agent. The compound supplying the positive ion and accepting the oxygen is called the reducing agent. Solids must be in a solution. Reactions can be explosive. Waste composition must be well known to prevent the inadvertent production of a more toxic or more hazardous end products. The reaction can be enhanced by catalysis, electrolysis, or irradiation. Reduction lowers the oxidation state of a compound. Reducing agents include iron, aluminum, zinc, and sodium compounds. For example, for cyanide removal by oxidation, the cyanide-bearing wastewater is oxidized with alkaline chlorine or hypochlorite solutions where in cyanide is initially oxidized to a less toxic cyanate and then to carbon dioxide and nitrogen.

$$Na\ CN + Cl_2 + 2NaOH \rightarrow NaCNO + 2\ NaCl + H_2O$$

$$2Na\ CNO + 3Cl_2 + 4NaOH \rightarrow 2CO_2 + N_2 + 6NaCl + 2H_2O$$

5.5.11 Chemical Oxidation

Oxidation destroys hazardous contaminants by converting them to non-hazardous or less toxic compounds that are stable, less mobile, or inert. The common oxidizing agents are ozone, hydrogen peroxide, hypochlorites, chlorine, and chlorine dioxide. Chemical oxidation depends on the client the oxidizing agents and the chemical contaminants. Chemical oxidation is also a part of the treatment process for cyanide-bearing and metals such as arsenic, iron, and manganese. Metal oxides formed in the oxidation process precipitate more readily out of the solution. Some compounds require a combination of oxidizing agents or the use of UV light with an oxidizing agent. Chemical oxidation increases the oxidation state of a contaminant and decreases the oxidation state of the reactant. The following is an example of an oxidation reaction (Table 5.5):

$$NaCN + H_2O_2 \rightarrow NaCNO + H_2O$$

5.5.12 Hydrolysis

Hydrolysis is the breaking of a bond in a nonwater-soluble molecule so that it will go into ionic solution with water. Hydrolysis can be achieved by adding chemicals, e.g., acid hydrolysis. It applies to a wide range of refractory organics. Hydrolysis is used to detoxify waste streams of carbamates, organophosphorous compounds, and other pesticides. Acid hydrolysis as an in situ treatment must be performed carefully due to potential mobilization of heavy metals. Depending on the waste stream, products may be unpredictable and the mass of toxic discharge may be greater than the waste originally input for treatment.

$$XY + H_2O \rightarrow HY + XOH$$

The process flow for a chemical oxidation system is as shown in Fig. 5.15.

Table 5.5 Effectiveness of chemical oxidation on general contaminant groups for liquids, soils, and sludge

Contaminant groups	Liquids	Soils/sludge
Organic		
Halogenated volatiles	O	♦
Halogenated semivolatiles	O	♦
Nonhalogenated volatiles	O	♦
Nonhalogenated semivolatiles	O	♦
PCBs	O	●
Pesticides	O	●
Dioxins/furans	♦	●
Organic cyanides	O	O
Organic corrosives	♦	♦
Inorganic		
Volatile metals	O	♦
Nonvolatile metals	O	♦
Asbestos	●	●
Radioactive materials	●	●
Inorganic corrosives	●	●
Inorganic cyanides	O	O
Reactive		
Oxidizers	●	●
Reducers	O	♦

O Demonstrated effectiveness: successful treatability test at some scale completed.
♦ Potential effectiveness: expert opinion that technology will work.
● No expected effectiveness: expert opinion that technology will not work.

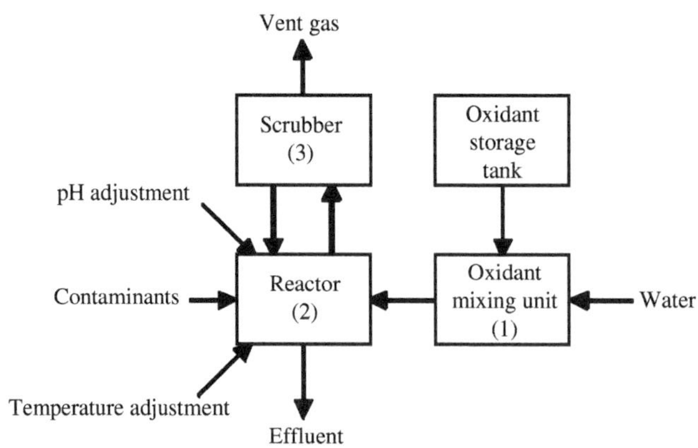

Figure 5.15 Process flow diagram for chemical oxidation system.

The reaction can be enhanced by adding UV light. Systems that combine ozone with hydrogen peroxide or UV radiation are catalytic ozonation processes. These accelerate ozone decomposition, increasing hydroxyl radical concentration, and promoting oxidation of the compounds. Solids must be in solution. Waste composition must be well known to prevent producing a more toxic or hazardous end product. Oxidation is not cost effective for highly concentrated waste because of the large amount of oxidizing agent required. Ozone can be used to pretreat wastes to break down refractory organics or to oxidize untreated organics after biological or other treatment processes. The limitations are the physical form of the waste (i.e., sludges and solids are not readily treated) and nonselective competition with other species. Ozonation systems have higher capital costs because ozone generators must be used. The cost of generating UV lights and the problems of scaling or coating on the lamps are two of the major drawbacks to UV-enhanced chemical oxidation systems. They do not perform well in turbid waters or slurries because reduced light transmission lowers the effectiveness.

5.6 CHEMICAL DEHALOGENATION

5.6.1 Alkaline Metal Dechlorination

This process of chemical dechlorination displaces chlorine from chlorinated organic compounds contained in oils and liquid wastes. Wastes are filtered before entering the reactor system and encountering the dechlorinating reagent. The great affinity of alkali metals for chlorine (or any halide) is the chemical basis of this process. The byproducts are chloride salts, polymers, and heavy metals. The reactor is blanketed with nitrogen because the reagents are sensitive to air and water, and an excess of reagent to chlorine is required. It is applied to such processes used to treat PCBs and others like chlorinated hydrocarbons, acids and dioxins. It has some limitations such as moisture content adversely affects the rate of reaction; therefore, dewatering should be a pretreatment step and waste stream concentrations are also important.

5.6.2 Electrolytic Oxidation

In electrolytic oxidation, cathodes and anodes are immersed in a tank containing waste to be oxidized, and a direct current is imposed on the system. This process is particularly applicable to cyanide bearing wastes. The reaction products are ammonia, urea, and carbon dioxide. During decomposition, metals are plated out on the cathodes. Electrolytic oxidation is used to treat high concentrations of up to 10% cyanide and to separate metals to allow their potential recovery. The limitations include the physical form of the feed (solids must be dissolved), nonselective competition with other species and long process times. Electrolytic recovery of single metal species can be 90% or higher.

5.6.3 Chemical Dehalogenation

Alkaline metal/polyethylene glycol (APEG) reagents effectively dechlorinate PCBs and oils, react rapidly to dehalogenate halo-organic compounds of all types. In the APEG reagents, alkali metal is held in solution by large polyethylene anions. PCBs and halogenated molecules are soluble in APEG reagents. These qualities combine in a single-phase system where the anions readily displace the halogen atoms. Halogenated aromatics react with APEGs resulting in the substitution of halogenated aromatics for chlorine atoms to form an APEG ether. The APEG ether decomposes to a phenol.

The APEG (waste preparation) involves a normal screening (1) for removing debris and large objects and producing particles that are small enough to allow the treatment in the reactor to avoid the binding of the mixer blades. Typically, reagent components are mixed with contaminated soil in the reactor (2). Treatment proceeds inefficiently without mixing. The mixture is heated to between 100°C and 180°C. The reaction proceeds for 1.5 h, depending upon the type, quantity, and concentration of the contaminants. The treated material goes from the reactor to a separator (3), where the reagent is removed and can be recycled (4). During the reaction, water is evaporated in the reactor, condensed (5), and collected for further use or recycled through the washing process, if required. Carbon filters (7) are used to trap any volatile organics that are not condensed. In the washer (6), soil is neutralized by the addition of acid. It is then dewatered (8) before disposal. Dehalogenation is effective in removing halogens from hazardous organic compounds such as dioxins, furans, PCBs, and chlorinated pesticides, rendering them nontoxic. Treatability tests should be conducted before the final selection of the APEG technology. Operating factors such as quantity of reagents, temperature, and treatment time should be defined. Treated soil may contain residual reagents and treatment byproducts that should be removed by washing the soil with water. The soil should also be neutralized by lowering the

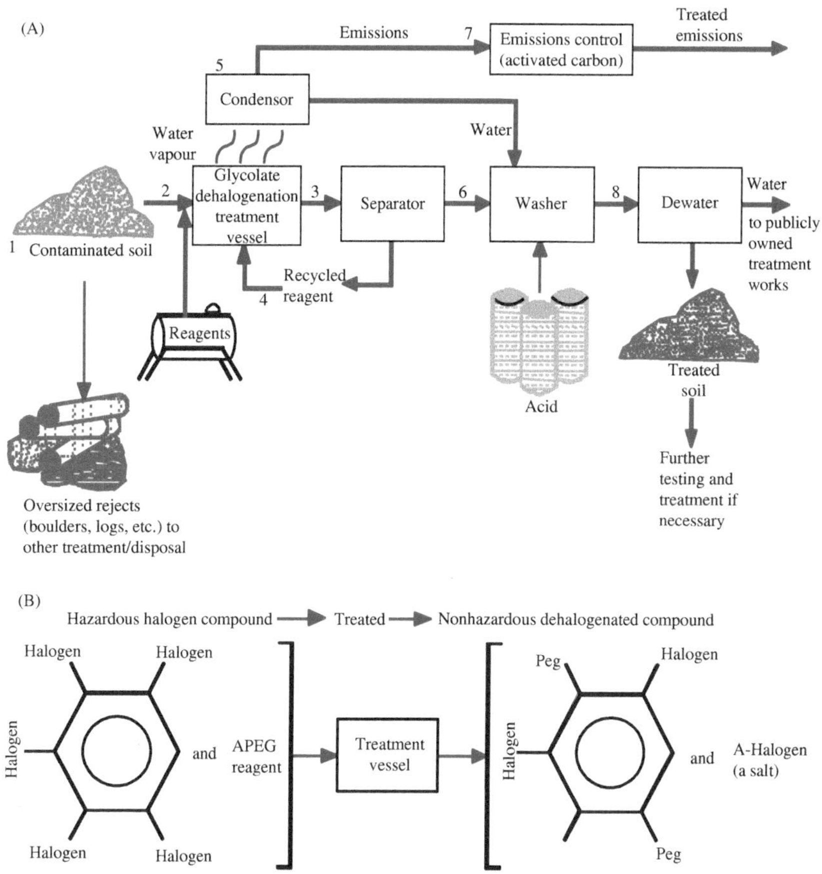

Figure 5.16 (A) Glycolate dehalogenation process flow. (B) Conceptual diagram of dehalogenation.

pH before final disposal. Specific safety aspects must be considered. Treatment of certain chlorinated aliphatics in high concentrations with APEG may produce potentially explosive compounds (e.g., chloroacetylenes) or cause a fire hazard. This process has been field tested (Fig. 5.16 and Table 5.6).

5.6.4 Chemical Dehalogenation: Based-Catalyzed Decomposition

Based-catalyzed decomposition (BCD) is another technology for removing chlorine molecules from organic substances.

Table 5.6 Effectiveness of APEG treatment on general contaminant groups for various matrices

Contaminant groups	Effectiveness sediments	Oil	Soil	Sludge
Organic				
Halogenated volatiles	★	★	★	★
Halogenated semivolatiles	★	★	★	★
Nonhalogenated volatiles	■	■	■	■
Nonhalogenated semivolatiles	■	■	■	■
PCBs	✓	✓	✓	✓
Pesticides (halogenated)	★	✓	✓	★
Dioxins/furans	✓	✓	✓	✓
Organic cyanides	■	■	■	■
Organic corrosives	■	■	■	■
Inorganic				
Volatile metals	■	■	■	■
Nonvolatile metals	■	■	■	■
Asbestos	■	■	■	■
Radioactive materials	■	■	■	■
Inorganic corrosives	■	■	■	■
Inorganic cyanides	■	■	■	■
Reactive				
Oxidizers	■	■	■	■
Reducers	■	■	■	■

✓ Demonstrated effectiveness: successful treatability test at some scale completed.
★ Potential effectiveness: expert opinion that technology will work.
■ No expected effectiveness: expert opinion that technology will not work.

Chlorinated Products + Catalyst, base, ↓
Dechlorinated products
+
Sodium chloride

The BCD process (Fig. 5.17) involves mixing the chemicals with the contaminated matrix (such as excavated soil or sediment or liquids containing these toxic compounds) and heating the mixture at 32–34°C for 1–3 hours. This process is applicable to PCB reduction from 4,000 ppm to less than 1 ppm. The BCD process requires only 1–5% reagent by weight. The reagent is also much less expensive than the APEG reagent. BCD also is regarded as effective for pentachlorophenol (PCP), PCBs,

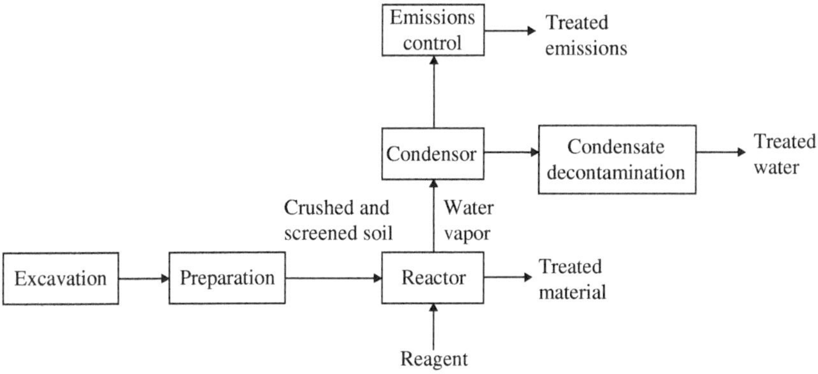

Figure 5.17 Based-catalyzed decomposition (BCD) process flow schematic.

pesticides (halogenated), herbicides (halogenated), dioxins, and furans. BCD is not intended as an in situ treatment. Treatability studies should be conducted before the final section. The off-gases are treated before releasing to the atmosphere. The treated receptor remains are nonhazardous, and can be either disposed of according to standard methods, or further processed to separate components for reuse.

5.6.4.1 Disposal of Hazardous Waste: Treatment Storage and Disposal Facility

General practice of the disposal of the hazardous waste is into the open drains, roadside dumps, landfills in low-lying areas, and municipal dustbins. It has to be disposed of scientifically at the place selected and approved by the State Pollution Control Boards/Pollution Control Committees (SPCBs/PCCs) after getting an environmental impact assessment of the place and having a public hearing. The hazardous waste should be disposed of at places specified and notified as TSDFs. A TSDF should have following attributes:

- It should be at least 0.5 km away from the inhabitations and proposed colonies.
- It should be away from the place of perennial supply of drinking water.
- It should have an access to transportation.
- It should be away from surface-water and ground-water accessibility to avoid contamination.
- It should not be susceptible to soil erosion and seismic activity.
- The site should be secured from easy access by the public.

- The site should be located close to a major waste generation area.
- The site should have enough facilities for disposing of the wastes.
- The waste should be treated before incinerated.
- Organic waste and wastes of chloride nature should not be burnt.
- PVC or plastic products should not be incinerated.
- The operator should have a license to conduct disposal of the waste.
- The operator should have the proper equipment, including autoclaving, incinerators and other equipment.

The integrated hazardous waste management facility should consist of a secured landfill, waste stores, incinerator, reuse/recycling facility, laboratory capable of comprehensive analysis, arrangement for transportation and handling of wastes including supporting infrastructure. Incineration of high-calorific-value hazardous wastes in cement kilns is a safe alternative to conventional disposal in dedicated waste incinerators. Based on the disposal pathway given by the laboratory after fingerprint analysis, the waste is subjected to pretreatment or stabilization, and taken to a landfill or for incineration.

5.6.4.1.1 Pretreatment or Stabilization of Hazardous Waste

Sometimes the properties of the waste do not permit it to go for landfilling or incineration directly, and it has to be pretreated; the process is termed as stabilization. Pretreatment is also required in the case of incineration in some situations. The basic principle of the stabilization is the immobilization or prevention of leachability of the toxic constituents. The methods adopted for stabilization of hazardous wastes are physical, chemical, biological, and thermal. The most commonly used stabilization processes are:

- *Immobilization*: Chemical binding of contaminants within a cementing structure to reduce the mobility or leachability of the waste constituents.
- *Encapsulation*: The occlusion or entrapment of containment particles within a solid matrix. The process is similar to cement sand mixture, where cement encapsulates sand. Encapsulation is the process where the wastes are enclosed within a stable water-resistant material. The encapsulated wastes must then be placed in a landfill or similar disposal site.
- *Solidification*: The conversion of slurries that do not readily dewater into solids by addition of solidification and adsorption agents. The in situ physical/chemical treatment, that is, solidification/stabilization, reduces the mobility of hazardous substances and contaminants in the

environment through both physical and chemical means. It seeks to trap or immobilize contaminants within their "host" medium (i.e., the soil, sand, and/or building materials that contain them) instead of removing them through chemical or physical treatment. In situ vitrification is another in situ solidification/stabilization process that uses an electric current to melt soil or other earthen materials at extremely high temperatures (1,600–2,000 °C or 2,900–3,650 °F) and thereby immobilize most inorganics and destroy organic pollutants by pyrolysis. The vitrification product is a chemically stable, leach-resistant, glass, and crystalline material similar to obsidian or basalt rock.

5.6.4.2 Limitations

1. Depth of contaminants may limit some types of application processes.
2. Future usage of the site may "weather" the materials and affect their ability to maintain immobilization of contaminants.
3. Some processes result in a significant increase in volume.
4. Reagent delivery and effective mixing are more difficult than for ex situ applications.
5. The solidified material may hinder future site use.
6. Processing of contamination below the water table may require dewatering.

The ex situ physical/chemical treatment (assuming excavation) has nine distinct innovative processes or groups of processes which include:

1. bituminization
2. emulsified asphalt
3. modified sulfur cement
4. polyethylene extrusion
5. Portland cement
6. radioactive waste solidification
7. sludge stabilization
8. soluble phosphates
9. molten glass

Applicability: The target contaminant group for ex situ solidification/stabilization is inorganics, including radionuclides. Most solidification/stabilization technologies have limited effectiveness against organics and pesticides, except vitrification, which destroys most organic contaminants.

5.6.4.3 Limitations

Environmental conditions may affect the long-term immobilization of contaminants. Some processes result in a significant increase in volume. Certain wastes are incompatible with different processes. Treatability studies are generally required. Organics are generally not immobilized. Long-term effectiveness has not been demonstrated for many contaminant/process combinations.

The process of stabilization is based on the comprehensive analysis report furnished by the waste generator, which gives a fairly clear idea if it requires stabilization or not. Then, laboratory scale studies have to be performed to arrive at an economically viable stabilization mechanism. Care must be taken in observing evolution of fumes, release of heat, release of ammonia gas, etc., and to the extent possible such combinations have to be avoided and necessary precautions to be followed while carrying out the stabilization process. The successful methodology has to be applied to the waste loads at the TSDF. The waste has to be thoroughly mixed with the stabilization reagents as per the protocol given by the laboratory. Water or leachate has to be added depending on the suitability and requirement. A sample has to be collected from the homogeneous mixture after the desired setting or curing period and poststabilization analysis has to be carried out to approve the effectiveness of the stabilization. The commonly used stabilization reagents and their properties are furnished below.

Lime: High pH forms hydroxides. It exhibits pozzolanic properties.

Cement: High pH, water absorbency, forms insoluble salts as $CaSO_4$ or $Ca_3(PO_4)_2$. It binds heavy metals.

Kiln dust: High pH, Ca forms insoluble salts as $CaSO_4$ or $Ca_3(PO_4)_2$, water absorbency and is a cheap reagent.

Rice husk: Fuel recovery, mixed with organic oils or solvents. It is used for better handling of slurry type organic sludges.

Rice husk ash: Activated carbon adsorbs organics.

Fly ash: Cementitious property and water absorbency. It is a cheap reagent.

Sawdust: High calorific value and absorbs organics.

Sodium silicate: Forms complexes with heavy metals, forms gel when mixed with water and hardens. It is expensive and handling is also a problem.

Bentonite: Absorbs organics. It is expensive reagent.

5.6.4.3.1 Secured Landfill

The concept of landfilling is to keep the waste in isolation from the environment. It is a containment system, which separates the waste from the surrounding environment. The objective is to mitigate the migration of leachate and minimization of emissions. The waste acceptance criteria for secured landfills is presented in Table 5.7.

In the construction of an engineered landfill the following liner systems are adopted depending on the national standards/strategies (Fig. 5.18).

- Mineral liner (clay lining)
- Geomembrane (HDPE liner)
- Drainage media
- Geotextile

Different countries have different standards of landfill designs, such as single-liner, double-liner, single-composite, and double-composite liner systems. The landfills can be above-ground or below-ground based on the ground-water profile and various other design parameters. For double-composite liner systems there should be at least one mineral liner (natural liner) and two synthetic liners (geomembranes) separated by a

Table 5.7 Concentration limits/criteria for acceptance of hazardous wastes for direct disposal in to secure landfill

Parameter	Limit
Physical	
Calorific value	Less than 3200 kcal/kg
LOI at 550°C	Less than 20% (nonbiodegradables) Less than 5% (biodegradables)
Flash point	More than 600°C
Paint-filter liquid test (PFLT)	Pass
Liquid release test	Pass
Chemical	
PH	Not less than 4 or greater than 11
Water-soluble organic substances	$\leq 10\%$
Water-soluble inorganic substances	$\leq 20\%$
Reactive cyanide	250 ppm
Reactive sulfide	500 ppm
TCLP levels	As per the list

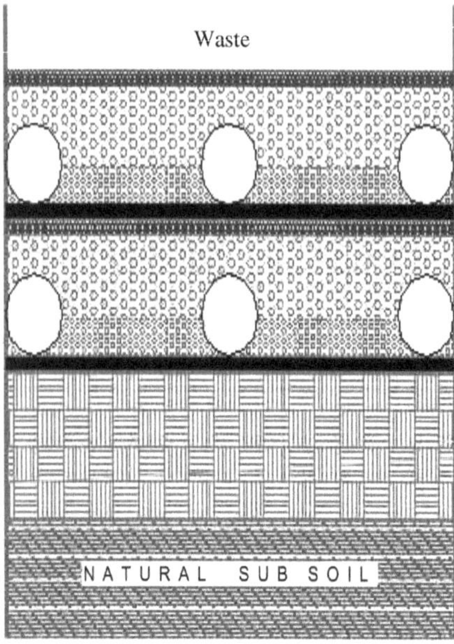

Figure 5.18 Cross-section of an engineered landfill's liner system.

drainage system, which acts as a leakage detection system. The wastes fit for direct disposal into the landfill and stabilized wastes will come to the landfill for disposal. The wastes have to be dumped in the landfill as per the advice given by the laboratory in the specified locations. The location of the grid has to be recorded, so that at a later date for any reason, the approximate location of the waste where it was deposited can be known. The waste has to be dozed, leveled, and compacted. At the end of every day's operations, daily soil has to be placed in about 15 cm thickness. The advantages of the daily cover are it achieves better compaction; ameliorates odors, dust emanation and gaseous emissions; improves esthetic appeal; and acts as a barrier between the layers and minimizes interreactions between the wastes, etc.

The leachate from the landfill can be collected from a vertical riser or side-slope riser. The leachate sump can be kept outside the landfill if it is an above-ground landfill. The active phase of the landfill should be as minimal as possible during the monsoon for the minimization of the leachate generation. It should be ensured that the hydraulic head on the geomembrane should not be more than 30 cm. Hence, the leachate should be pumped out as soon as it is generated. For the intermediate cover during the monsoon closure, soil of about 30 cm to be placed. Whereas for the final cover it is about 45–60 cm to be placed as a topsoil requirement for the vegetation to develop. While planning and designing the common hazardous waste landfill the following criteria needs to be taken into consideration:

- a liner system at the base and sides to prevent migration of leachate or gas to the surrounding soil.
- a leachate collection and treatment facility;
- a gas collection and treatment facility (optional);
- a final cover system at the top of the landfill, which enhances surface drainage;
- prevents infiltration of water and supports surface vegetation;
- a surface-water drainage system, which collects and removes all surface runoff from the landfill site;
- an environmental monitoring system which periodically collects and analyses air, surface water, soil-gas, and ground-water samples around the landfill site;
- a closure and postclosure plan that lists the steps that must be taken to close and secure the landfill site.

The liner system for leachate control within a landfill involves prevention of migration of leachates from landfill sides and base to the subsoil and drainage of leachate collected at the base of a landfill.

5.6.4.3.2 Incineration

Incineration is applied to wastes that cannot be recycled, reused, or safely deposited in a landfill. It is a high-temperature, thermal oxidation process in which hazardous wastes are converted in the presence of oxygen in the air into gases and incombustible solid residue. Gases are vented into the atmosphere through a gas-cleaning system and solid residue goes to a landfill. The applicability of incineration of hazardous waste depends on certain considerations, such as if the waste is biologically hazardous, it is resistant to biodegradation and persistent, volatile, and therefore easily dispersed, and cannot be safely disposed in a landfill even after stabilization and volume reduction. The typical wastes that would need to be incinerated may include solvent wastes (spent solvents), waste oils, oil emulsions, oil mixtures, pesticide wastes, pharmaceutical wastes, refinery wastes, phenolic wastes, grease and wax wastes, organic wastes (containing halogens, sulfur, phosphorous or nitrogen compounds) and material contaminated with oil and others with calorific value >2500 kcal/kg. Incineration aims are destroying the toxicity of wastes and yield products of combustion that are harmless. The temperature, time, and turbulence are important parameters for combustion. Availability of oxygen is additional parameter that forms an integral part of the incineration system. When waste is burnt at a higher temperature, destruction would be complete and formulation of unburnt waste, formation of organic byproducts, etc. would be eliminated. The longer the waste is held at high temperature, the greater will be the degree of destruction and the less likelihood of formation products of incomplete combustion. Greater turbulence relates to the degree of mixing between the waste and oxygen in the combustion air and to the absence of temperature gradients within the furnace. Greater turbulence provides better control, better access to air and more complete oxidation, and destruction of waste being burnt. Temperatures of 900–1100°C can be expected for hydrocarbon wastes and 1100–1200°C for certain wastes like PCBs, waste oil residues, etc. For other halogenated organics, a case-by-case approach may be needed. A minimum gas phase residence time of 2 seconds has to be maintained. Combustion air is in 100% excess of stoichiometric requirements. Turbulence is achieved through good incinerator design. The destruction and removal efficiency (DRE %) is as follows:

$$DRE = \left[(W_{\text{in}} - W_{\text{out}})/W_{\text{in}}\right] \times 100\%$$

where

W_{in} is the concentration of the compound in the waste feed × mass rate of feed of principal organic hazardous constituent (POHC)

W_{out} is the concentration of the compound in the stack gas × volumetric flow rate of stack gas

As a rule, destruction removal efficiency (DRE) must be greater than 99.9%. The gas cleaning is to remove, as completely as practicable, particulates and noncombustible contaminants such as fly ash and metal oxides and acidic gases (particularly HCl). Unburnt wastes and tract organic byproducts are also required to be removed by the gas-cleaning equipment. Air-pollution control for incineration of hazardous waste is required to meet emission standards. In the Alstom design, the gases are first cooled in a quench tower. Activated carbon and lime are sprayed into the cooled gases and led to a pulsed jet-bag filter followed by an alkaline scrubber and then to a stack. This shall meet the emissions guidelines for incinerators. Fuel blending of organic waste and thermal treatment at lower temperatures in certain cases can be a low-cost option for waste disposal as an alternate to incineration; however, the emissions standards applicable for these would be the same as applicable to incinerators (Fig. 5.19).

Figure 5.19 A large capacity hazardous waste incinerator.

5.7 MONITORING PROTOCOL FOR TREATMENT STORAGE AND DISPOSAL FACILITY

5.7.1 Ambient Air-Quality Monitoring

Air-quality monitoring stations at upwind, downwind, and at three stations at a 120 degree angle around the TSDF are necessary. The locations of air-quality monitoring stations depend on the stack height and location of any particular ecologically sensitive feature around the disposal facility. Apart from standard parameters under the National Ambient Air Quality Standards (NAAQS), additional parameters, namely, total volatile organic compounds (VOCs) and polycyclic aromatic hydrocarbons (PAHs) to align the monitoring program with the potential impacts of TSDF operations, should be monitored. The parameters, namely, SPM, RSPM, NO_x, and SO_x, should be continued to be monitored as per National Ambient Air Quality Standards (NAAQS0 criteria (minimum of 104 measurements in a year taken twice a week, 24 hourly). In addition, VOCs (total) and PAHs should be monitored at least twice in a year (premonsoon and post-monsoon). The parameters to be monitored and the frequency stack gaseous emission from the incinerator are SO_2, NO_x, HCI, and CO. Suggested parameters and the frequency of monitoring of the vent gases attached with the capped Secured Land-Fill (SLF) are total VOCs, and

H_2S should be monitored at least once in a month through the vents of the capped cells till designed lifespan of the TSDF.

Air-quality testing in terms of biological indicator includes plantations of locally available sensitive plants to be planted in all directions of the TSDF and at different distances, and to observe and record periodically the health of each plant. The air pollution tolerance index (APTI) should be observed.

5.7.2 Ground-Water Quality Monitoring

It is recommended to monitor ground-water characteristics at least once a quarter throughout the designated lifespan of the TSDF. Parameters to be analyzed/recommended are pH, color, electrical conductivity, turbidity (NTU), total suspended solids, total dissolved solids, total organic carbon, chemical oxygen demand, chlorides, nitrates, sulfates, total Kjeldahl nitrogen, total alkalinity, total hardness and pesticides, and heavy metals (such as Pb, Cd, Cu, Zn, Cr, Hg, Ni, Fe, CN, As, and Mn). It is recommended that the ground-water samples should be collected at least up to a distance of 5 km from the TSDF location. If no open wells or tube wells are available, action needs to be taken to provide at least four monitoring wells (piezometric) around the TSDF, i.e., one on the up-gradient of the ground-water flow and other three on the down-gradient side of the ground-water flow at least up to the first-layer aquifer. Depending upon the situation, if required, the monitoring well till the second aquifer should also be extended.

The directions of the ground-water flow have to be established in consultation with the state ground-water board or any other authority. The ground-water flow direction has to be follow local conditions such as draw down of ground water.

5.7.3 Surface-Water Quality Monitoring

Monitoring of surface waters (nullah/river, impoundments) of upstream and downstream and in adjoining area is necessary at least once in a quarter. It is also necessary to collect a sample of benthos deposits from the stream up to a distance of 500 m from the TSDF. It is recommended that the surface-water samples should be analyzed for pH, color, electrical conductivity, turbidity (NTU), suspended solids, total dissolved solids, total organic carbon, dissolved oxygen, biochemical oxygen demand, chemical oxygen demand, chlorides, nitrates, sulfates, total Kjeldahl nitrogen, total alkalinity, total hardness, and heavy metals such as Pb, Cd, Cu, Zn, Cr, Hg, Ni, Fe, CN, As, and Mn.

Figure 5.20 Treatment storage and disposal facility (TSDF). *Courtesy: Ramky Group.*

5.7.4 Soil-Quality Monitoring

Soil samples to be monitored and recommended parameters are pH, electrical conductivity, color, total dissolved solids, total organic carbon, PAH, and heavy metal such as Pb, Cd, Cu, Zn, Cr, Hg, Ni, cyanide, As, and Mn. At least one number of composite soil sample is required to be collected up to a depth of 1 m beneath the soil surface for every grid size of 250 × 250 m up to a radius of 500 m from the center of the TSDF. It is recommended that the soil samples should be collected and analyzed for the suggested parameters at least once in a year, i.e., premonsoon (Fig. 5.20).

5.8 LEGAL PROVISION

The Hazardous Wastes (Management and Handling) Rules, 1989, and amendments, and the Hazardous Wastes (Management Handling and Transboundary Movement) Rules, 2009, and amendments under the Environment (Protection) Act, 1986, govern the management of hazardous waste. However, the legal provisions change from time to time depending on the minor amendments made in the Rules for handling and management of wastes.

CHAPTER 6

Electronic Waste

6.1 INTRODUCTION TO ELECTRONIC WASTE

The electronic industry is the largest and fastest-growing manufacturing industry. All over the world there is a tremendous growth in the field of information technology. Since the 1990s there has been strong growth in the electronic equipment market among which computers, mobile phones, televisions, printers, refrigerators, and washing machines have shown the strongest growth. Mobile phones are not just tools of communication but have taken a new meaning altogether. They are used as indoor video games, cameras, video recorders, computers, and telecommunication devices. With the number of mobile phone users increasing, the problem of safe disposal of discarded mobile phones is becoming a troublesome task. E-waste is one of the fastest-growing waste streams in India due to increasing market penetration in developing countries, replacement market in developed countries, and high obsolescence rate.

All electronic and electrical items, on completion of their useful life, are being discarded rapidly and contribute to the huge quantum of e-waste. The generation of e-waste has grown manyfold since the 1990s and will continue to accelerate at a fast pace. The National Safety Council estimates almost 100 million computers and monitors become obsolete annually. The present scientific disposal system processes only 15–20% of the total e-waste generated. The balance is being disposed of by unauthorized recyclers or thrown out into garbage dumps. The growing dependence on IT and electronic products has given rise to new environmental challenges.

E-waste dismantling or incineration is considered toxic. This waste is targeted for reuse, recovery, or hazardous waste disposal. The recovery of metals is a profitable business, which results in local, transboundary, and global trade. Environmental issues and trade associated with e-waste has driven the definition of e-waste both at national and international level. In this context, it is important to understand the existing waste electrical and electronic equipment (WEEE)/e-Waste, its emerging problem, and also its potential as a business opportunity of increasing significance, given

Solid and Hazardous Waste Management.
DOI: http://dx.doi.org/10.1016/B978-0-12-809734-2.00006-7

the volumes of e-waste being generated and the content of both toxic and valuable materials in them.

The market for e-waste in India is not concentrated in a single place, but spread over different areas, each handling a different aspect of recycling. No sophisticated machinery or personal protective equipment (PPE) is used for the extraction of different materials. All the work is done by bare hands and only with the help of hammers and screwdrivers. Children and women are routinely involved in the operations.

Electronic waste (e-waste) comprises waste electronics/electrical goods that are not fit for their original intended use or have reached their end of life. According to the guidelines on e-waste management brought up by Ministry of Environment and Forests (MoEF), Government of India, e-waste is such waste comprising of wastes generated from used electronic devices and household appliances that are not fit for their original intended use and are destined for recovery, recycling, or disposal. Such wastes encompass a wide range of electrical and electronic devices such as computers, handheld cellular phones, personal stereos, and large household appliances such as refrigerators, air conditioners, etc.

Public perception of e-waste is often restricted to a narrower sense, comprising mainly of end-of-life information and telecommunication equipment and consumer electronics. However, technically, e-waste is only a subset of WEEE. According to the Organization for Economic Cooperation and Development (OECD) any appliance using an electric power supply that has reached its end-of-life would come under WEEE. Globally, WEEE/e-waste is the most commonly used terms for e-waste. There is no standard definition of WEEE/e-waste. A number of countries have come out with their own definitions, interpretations, and usage of the term e-waste/WEEE.

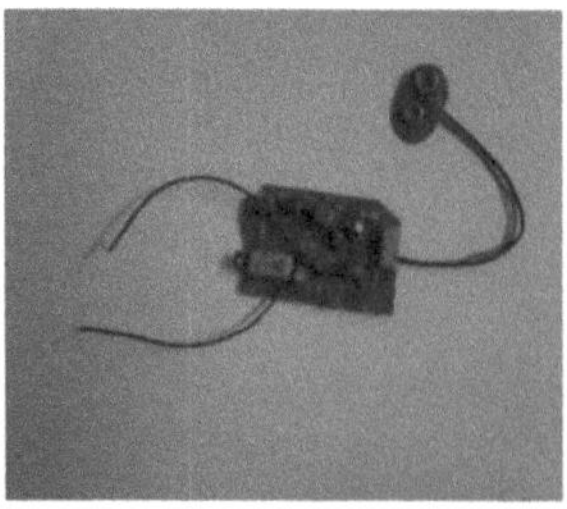

The term e-waste refers to the technological characteristics and the hazardous materials incorporated in this waste. It refers to a broad range of electric and electronic goods that have outlived their use for producers and consumers, are ready for disposal, and that contain chemical materials considered hazardous for humans and for our natural environment.

The software sector of the information technology industry has led to a prominent global presence of India. Policy changes have led to a tremendous influx of leading multinational companies into India to set up manufacturing facilities, research and development centers, and software development facilities. The domestic market is getting revitalized due to buoyant economic growth and changing consumption patterns. This growth has significant economic and social impacts. The increase of electronic products, consumption rates, and higher obsolescence rate leads to higher generation of e-waste. The increasing obsolescence rates of electronic products added to the huge import of junk electronics from abroad create a complex scenario for solid waste management in India. Among the electronic items, computers, televisions, and mobile phones are finding their way into the waste stream. The various components/parts of these are as follows:

- *computers*: motherboard, switch mode power supply, random access memory, hard disk, processors, capacitors, integrated circuits, main board, magnetic touching sheet, CD drive, floppy drive, and diodes, etc.
- *televisions*: capacitors, resistors, transformers, supply transformer regulator, integrated circuits, line output transformer, tuners, condensers, socket, color picture tube, Zener diode, and normal diode, etc.
- *mobile phones*: lens, internal antenna, aerial, speakers, earpiece, microphone, microphone connectors, loud speakers, buzzers, ringers, charging blocks, system connectors, chassis, slide mechanism, ribbon cables, SIM slot covers, readers, backup, battery, battery clip, covers, battery contacts, connectors, keypad membrane, etc.

6.2 AVERAGE LIFE OF ELECTRONIC GOODS/AVERAGE LIFE CYCLE/OBSOLESCENCE RATE

Average life of electronic goods/average life cycle/obsolescence rate is the time span after which the electrical and electronic item comes to its "end of life." It can be defined in terms of active life, passive life, and storage.

Average life cycle/Obsolescence rate = Active Life + Passive Life + Storage

The number of years a machine can be effectively used is called its active life. After active life, it can be refurbished or reused for certain time period. This time period constitutes passive life. Storage includes storage time before disposal and storage at repair shops before dismantling.

In developed countries, average life cycle of electrical and electronic equipment is generally equivalent to active life, while in developing countries, it is a sum of active life, passive life, and storage. Therefore, in developing countries, a second-hand market exists for WEEE/e-waste after its active life. All three parameters vary in different geographical regions. Therefore, the average life cycle/obsolescence rate varies in each geographical region and leads to different WEEE/e-waste inventory. Electronic equipment/products become obsolete due to advancement in technology; changes in fashion, style, and status; and as they near the end of their useful life and quality of power supply.

6.3 CLASSIFICATION OF E-WASTE

E-waste can be classified as computer peripherals, telecommunication devices, and industrial electronics. The details are as follows:

- computer peripherals: monitor, keyboard, mouse, circuit boards, CDs, floppies, laptops, servers, etc.
- telecommunication devices: phones, cell phones, pagers, fax machine, routers, transmitters, Radio Frequency (RF) equipment, etc.
- industrial electronics: sensors, automobile electronic devices, medical devices, etc.
- lighting devices: fluorescent tubes
- household appliances: TV, fridge, washing machine, video, camera, etc.

6.4 CATEGORIES OF E-WASTE

As per the draft E-Waste (Management and Handling) Rules, there are nine categories of e-waste. They are large household appliances, small household appliances, toys, leisure and sports equipment, electrical and electronic tools, medical devices, monitoring and control instruments, automatic dispensers, information technology (IT) and telecommunication equipment, and consumer electronics. The list of products covered under the categories is as given in Table 6.1.

Table 6.1 List of products covered under the categories

Category	Products
Large household appliances	Refrigerators and freezers; other appliances used for refrigeration, conservation, and storage of food; washing machines; clothes dryers; dishwashing machines; cooking ranges/stoves; electric hot plates; microwaves; other appliances used for cooking and other processing of food; electric heating appliances; electric radiators; other fanning, exhaust, ventilation, and conditioning equipment
Small household appliances	Vacuum cleaners, carpet sweepers; other appliances used for cleaning, appliances used for sewing, knitting, weaving, and other processing for textiles; irons and other appliances used for ironing and other care of clothing; toasters, fryers, grinders, coffee machines, and equipment for opening or sealing containers or packages; electric knives; appliances for haircutting, hair drying, tooth brushing, shaving, massage, and other body care appliances; digital clocks, watches, and equipment for the purpose of measuring indicating or registering time scales
Toys, leisure and sports equipment	Electric trains or car-racing sets; handheld video game consoles; video games; computers for biking, diving, running, rowing, etc.; sports equipment with electric or electronic components; coin slot machines
Electrical and electronic tools (except large-scale stationary industrial tools)	Drills; saws; sewing machines; equipment for turning, milling, sanding, grinding, sawing, cutting, shearing, drilling, making holes, punching, folding, bending, or similar processing of wood, metal, and other materials; tools for

(Continued)

Table 6.1 (Continued)

Category	Products
	riveting, nailing, or screwing or removing rivets, nails, screws, or similar uses; tools for welding, soldering, or similar use equipment for spraying, spreading, dispersing, or other treatment of liquid or gaseous substances by other means; tools for mowing or other gardening activities
Medical devices (except implanted and infected products)	Radiotherapy equipment; cardiology; dialysis; pulmonary ventilators nuclear medicine; laboratory equipment for in-vitro diagnosis; analyzers; freezers; fertilization tests; other appliances for detecting, preventing, monitoring, treating, alleviating illness, injury, or disability
Monitoring and control instruments	Smoke detectors; heating regulators; thermostats; measuring, weighing, or adjusting appliances for household or as laboratory equipment; other monitoring and control instruments used in industrial installations (e.g., in control panels)
Automatic dispensers	Automatic dispensers for beverages, automatic dispensers for hot or cold bottles or cans, automatic dispensers for solid products, automatic dispensers for money, all appliances which deliver automatically all kind of products
IT and telecommunication equipment	Centralized data processing; mainframes; minicomputers; personal computing; personal computers (CPU with input and output devices); laptops (CPU with input and output devices); notebooks, notepads, etc.; printers; copying equipment; electrical and electronic typewriters; pocket and desk calculators; other products and equipment for the collection,

(*Continued*)

Table 6.1 (Continued)

Category	Products
	storage, processing, presentation, or communication of information by electronic means; user terminals and systems; facsimiles; telexes; telephones; pay telephones; cordless telephones; cellular telephones; answering systems; and other products or equipment of transmitting sound, images, or other information by telecommunications
Consumer electronics	radio sets; television sets; video cameras; video recorders; digital cameras; hi-fi recorders; audio amplifiers; musical instruments and other products or equipment for the purpose of recording or reproducing sound or image, including signals or other technologies for the distribution of sound and image than by telecommunications

As per the E-Waste (Management and Handling) Rules, 2011, e-waste has been categorized into two categories, namely (1) electrical and electronic equipment and (2) consumer electrical and electronics. The categories of electrical and electronic equipment are as given in Table 6.2.

6.5 COMPOSITION AND CHARACTERISTICS OF E-WASTE

The composition of e-waste is very diverse and differs in products across different categories. It contains more than 1000 different substances, which fall into two categories: hazardous and nonhazardous. Broadly, e-waste consists of ferrous and nonferrous metals, plastics, glass, wood, and plywood, circuit boards, concrete and ceramics, rubber, and other items. Iron and steel constitutes about 50% of the e-waste followed by plastics (21%), nonferrous metals (13%), and other constituents 16%. Nonferrous metals consist of metals like copper, aluminum, and precious metals like silver, gold, platinum, palladium, etc. The presence of elements like lead, mercury, arsenic, cadmium, selenium, and hexavalent chromium and

Table 6.2 Categories of electrical and electronic equipment

S. No.	Category	List of products
1.	Information technology and telecommunication equipment	Centralized data processing: mainframes, minicomputers Personal computing: personal computers (central processing unit with input and output devices), laptop computers (central processing with input and output devices), notebook computers, notepad computers Printers including cartridges, copying equipment, electrical and electronic typewriters, user terminals and systems, facsimile, telex, telephones, pay telephones, cordless telephones, cellular telephones, and answering systems
2.	Consumer electrical and electronics	Television sets including sets based on (liquid crystal display and light-emitting diode technology), refrigerator, washing machine, air conditioners excluding centralized air conditioning plants.

flame retardants beyond threshold quantities classifies them as hazardous waste. The various parts/materials/ composition of e-waste may be divided broadly into six categories:

1. iron and steel, used for casings and frames
2. nonferrous metals, especially copper used in cables, and aluminum
3. glass used for screens, windows
4. plastic used as casing, in cables, and for circuit boards
5. electronic components
6. others (rubber, wood, ceramic, etc.).

E-waste is much more hazardous than any other municipal wastes because electronic gadgets contain thousands of components made of deadly chemicals and metals like lead, cadmium, chromium, mercury, polyvinyl chlorides (PVCs), brominated flame retardants, beryllium, antimony, and phthalates. Long-term exposure to these substances damages the nervous system, kidneys, bones, and reproductive and endocrine systems. Some of these substance are carcinogenic and neurotoxic. A study conducted by Greenpeace in 2005 in electronic recycling yards in Delhi indicated that the presence of high levels of hazardous chemicals including dioxins and furans in the areas where this primitive/unauthorized recycling takes place.

Disposal of e-wastes is a critical problem and poses a threat to both health and vital components of the ecosystem. There are a number of channels through which ewaste goes to the environment. E-waste that is landfilled produces contaminated leachates, which eventually pollute the ground water. Acids and sludge obtained from melting computer chips, if disposed on the ground, cause acidification of soil, leading to contamination of water resources. Incineration of e-wastes can emit toxic fumes and gases, thereby polluting the surrounding air. Improper recycling and recovery methods can have major impacts on the environment. Crude forms of dismantling can often lead to toxic emissions, which pollute the air and thereby also expose the workers to harmful materials. The most dangerous form of recycling and recovery from e-waste is the open-burning of circuit boards (made of plastic) in order to recover copper and other metals. Extraction of metals through acid-bath method or through mercury amalgamation also contributes to environmental degradation.

The toxic materials present in equipment can be an environmental as well as a health hazard. Mercury will leach when certain electronic devices, such as circuit breakers are destroyed. Not only does the leaching of mercury pose problems, the vaporization of metallic mercury and dimethylene mercury is also of concern. The same is true for polychlorinated biphenyls (PCBs) from condensers. When brominated flame retardant plastic or cadmium-containing plastics are landfilled, both polybrominated diphenyl ethers (PBDEs) and cadmium may leach into the soil and ground water. It has been found that significant amounts of lead are dissolved from broken lead-containing glass, such as the cone glass of cathode ray tubes (CRTs), which gets mixed with acid waters; this is a common occurrence in landfills.

The rapid growth and faster change in modules of computers, cell phones, and consumer electronics becomes a major issue that increases the amount of ewaste generation.

6.5.1 Hazardous Substances: Their Occurrences and Impact on Environmental and Human Health

Hazardous substances, and their occurrences and impact on environmental and human health, are as given in Table 6.3.

Table 6.3 Hazardous substances: their occurrence and impacts on environmental and human health

Substance	Occurrence in e-waste	Environmental and health relevance
PCB (polychlorinated biphenyls)	Condensers, transformers	Cause cancer, effects on the immune system, reproductive system, nervous system, endocrine system, and other health effects. Persistent and bioaccumulative
TBBA (tetrabromobisphenol-A) • Polybrominated biphenyls (PBB) • Polybrominated diphenyl ethers (PBDE)	Fire retardants for plastics (thermoplastic components, cable insulation) TBBA is presently the most widely used flame retardant in printed wiring boards and covers for components	Can cause long-term period injuries to health; acutely poisonous when burned
Chlorofluorocarbon (CFC)	Cooling unit, insulation foam	Combustion of halogenated substances may cause toxic emissions
Polyvinyl chloride (PVC)	Cable insulation	High-temperature processing of cables may release chlorine, which is converted to dioxins and furans
Arsenic	Small quantities in the form of gallium arsenide within light-emitting diodes	Acutely poisonous and on a long-term perspective injurious to health
Barium	Getters in CRT	May develop explosive gases (hydrogen) if wetted

(Continued)

Table 6.3 (Continued)

Substance	Occurrence in e-waste	Environmental and health relevance
Beryllium	Power-supply boxes which contain silicon-controlled rectifiers, beamline components	Harmful if inhaled
Cadmium	Rechargeable Ni-Cd batteries, fluorescent layer (CRT screens), printer inks, and toners	Acutely poisonous and injurious to health on a long-term perspective
Chromium VI	Data tapes, floppy disks	Acutely poisonous and injurious to health on a long-term perspective causes allergic reactions
Gallium arsenide	Light-emitting diode (LED)	Injurious to health
Lithium	Li-batteries	May develop explosive gases (hydrogen) if wetted
Hazardous Substances: their occurrence and impacts on environmental and human health		
Mercury	Found in the fluorescent lamps that provide backlighting in LCDs, in some alkaline batteries and mercury wetted switches	Acutely poisonous and injurious to health on a long-term perspective
Nickel	Rechargeable Ni–Cd-batteries or NiMH batteries, electron gun in CRT	May cause allergic reactions
Rare earth elements (yttrium, europium)	Fluorescent layer (CRT screen)	Irritates skin and eyes

(Continued)

Table 6.3 (Continued)

Substance	Occurrence in e-waste	Environmental and health relevance
Zinc sulfide	Used on the interior of a CRT screen, mixed with rare earth metals	Toxic when inhaled
Toxic organic substances	Condensers, liquid crystal display	
Toner dust	Toner cartridges for laser printers/copiers	Health risk when dust is inhaled risk of explosion

Source: Report on assessment of electronic wastes in Mumbai–Pune area—MPCB, March 2007.

The effect on health and the hazardous nature of a few of the constituents of e-waste is as follows.

6.5.2 Arsenic

Arsenic is a poisonous metallic element that is present in dust and soluble substances. Chronic exposure to arsenic can lead to various diseases of the skin and decrease nerve conduction velocity. Chronic exposure to arsenic can also cause lung cancer and can often be fatal.

6.5.3 Barium

Barium is a metallic element that is used in sparkplugs, fluorescent lamps, and getters in vacuum tubes. Being highly unstable in the pure form, it forms poisonous oxides when in contact with air. Short-term exposure to barium could lead to brain swelling, muscle weakness, and damage to the heart, liver, and spleen. Animal studies reveal increased blood pressure and changes in the heart from ingesting barium over a long period of time. The long-term effects of chronic barium exposure to human beings are still not known due to lack of data.

6.5.4 Beryllium

Beryllium has recently been classified as a human carcinogen because exposure to it can cause lung cancer. The primary health concern is inhalation of beryllium dust, fumes, or mist. Workers who are constantly exposed to beryllium, even in small amounts, and who become sensitized to it can develop what is known as chronic beryllium disease (beryllicosis), a disease that primarily affects the lungs. Exposure to beryllium also causes a form of skin disease that is characterized by poor wound healing and wartlike bumps. Studies have shown that people can still develop beryllium diseases even many years following the last exposure.

6.5.5 Brominated Flame Retardants

The three main types of brominated flame retardants (BFRs) used in electronic and electrical appliances are polybrominated biphenyl (PBB), polybrominated diphenyl ether (PBDE), and tetrabromobisphenol-A (TBBPA). Flame retardants make materials, especially plastics and textiles, more flame resistant. They have been found in indoor dust and air through migration and evaporation from plastics. Combustion of halogenated case material and printed wiring boards at lower temperatures releases toxic emissions including dioxins, which can lead to severe hormonal disorders. Major electronics manufacturers have begun to phase out brominated flame retardants because of their toxicity.

6.5.6 Cadmium

Cadmium components may have serious impacts on the kidneys. Cadmium is adsorbed through respiration but is also taken up with food. Due to its long half-life in the body, cadmium can easily be accumulated in amounts that cause symptoms of poisoning. Cadmium shows a danger of cumulative effects in the environment due to its acute and chronic toxicity. Acute exposure to cadmium fumes causes flulike symptoms of weakness, fever, headache, chills, sweating, and muscular pain. The primary health risks of long-term exposure are lung cancer and kidney damage. Cadmium is also believed to cause pulmonary emphysema and bone disease (osteomalacia and osteoporosis). http://www.intox.org/databank/documents/chemical/cadmium/ehc135.htm

6.5.7 Chlorofluorocarbons

Chlorofluorocarbons (CFCs) are compounds composed of carbon, fluorine, chlorine, and sometimes hydrogen. Used mainly in cooling units and insulation foam, they have been phased out because when released into the atmosphere, they accumulate in the stratosphere and have a deleterious effect on the ozone layer. This results in increased incidence of skin cancer in humans and in genetic damage in many organisms.

6.5.8 Chromium

Chromium and its oxides are widely used because of their anticorrosive properties. While some forms of chromium are nontoxic, chromium (VI) is easily absorbed in the human body and can produce various toxic effects within cells. Most chromium (VI) compounds are irritating to eyes, skin, and mucous membranes. Chronic exposure to chromium (VI) compounds can cause permanent eye injury, unless properly treated. Chromium VI may also cause DNA damage.

6.5.9 Dioxins

Dioxins and furans are a family of chemicals comprising 75 different types of dioxin compounds and 135 related compounds known as furans. The term dioxins is taken to mean the family of compounds comprising polychlorinated dibenzo-*p*-dioxins (PCDDs) and polychlorinated dibenzofurans (PCDFs). Dioxins have never been intentionally manufactured, but form as unwanted byproducts in the manufacture of substances like some pesticides as well as during combustion. Dioxins are known to be highly toxic to animals and humans because they bioaccumulate in the body and can lead to malformations of the fetus, decreased reproduction and growth rates, and cause impairment of the immune system among other things. The best-known and most toxic dioxin is 2,3,7,8-tetrachlorodibenzo-*p*-dioxin (TCDD).

6.5.10 Lead

Lead is the fifth most widely used metal after iron, aluminum, copper, and zinc. It is commonly used in the electrical and electronics industry in solder, lead-acid batteries, electronic components, cable sheathing, in the glass of CRTs, etc. Short-term exposure to high levels of lead can cause vomiting, diarrhea, convulsions, coma, or even death. Other symptoms are appetite loss, abdominal pain, constipation, fatigue, sleeplessness, irritability, and headache. Continued excessive exposure, as in an industrial setting, can affect the kidneys. It is particularly dangerous for young children because it can damage nervous connections and cause blood and brain disorders.

6.5.11 Mercury

Mercury is one of the most toxic yet widely used metals in the production of electrical and electronic applications. It is a toxic heavy metal that bioaccumulates causing brain and liver damage if ingested or inhaled. In electronics and electrical appliances, mercury is highly concentrated in batteries, some switches and thermostats, and fluorescent lamps.

6.5.12 Polychlorinated biphenyls

PCBs are a class of organic compounds used in a variety of applications, including dielectric fluids for capacitors and transformers, heat-transfer fluids, and as additives in adhesives and plastics. PCBs have been shown to cause cancer in animals. PCBs have also been shown to cause a number of serious noncancer health effects in animals, including effects on the immune system, reproductive system, nervous system, endocrine system, and other health effects. PCBs are persistent contaminants in the environment. Due to the high lipid solubility and slow metabolism rate of these chemicals, PCBs accumulate in the fat-rich tissues of almost all organisms (bioaccumulation). The use of PCBs is prohibited in OECD countries, however, because due to its wide use in the past, it still can be found in waste electrical and electronic equipment as well as in some other wastes.

6.5.13 Polyvinyl Chloride

PVC is the most widely used plastic, used in everyday electronics and appliances, household items, pipes, etc. PVC is hazardous because it contains up to 56% chlorine, which when burned produces large quantities of hydrogen chloride gas, which combines with water to form hydrochloric acid. It is dangerous because when inhaled, it leads to respiratory problems.

6.5.14 Selenium

Exposure to high concentrations of selenium compounds cause selenosis. The major signs of selenosis are hair loss, nail brittleness, and neurological abnormalities (such as numbness and other odd sensations in the extremities).

6.5.15 Substances and Elements in E-Waste

The list of substances and elements contained in e-waste in bulk and in small quantities are as follows.

6.5.15.1 Substances in Bulk in E-Waste

The substances in bulk in e-waste in alphabetical order are epoxy resins, fiberglass, PCBs, PVC, and thermosetting plastics.

6.5.15.2 Elements in Bulk in E-Waste

The elements in bulk in e-waste are lead, tin, copper, silicon, beryllium, carbon, iron, and aluminum.

6.5.15.3 Elements in Small Amounts in E-Waste

Elements in small amounts in e-waste are cadmium, mercury, and thallium.

6.5.15.4 Elements in Trace Amounts in E-Waste

Elements in trace amounts in e-waste are (alphabetically) americium, antimony, arsenic, barium, bismuth, boron, cobalt, europium, gallium, germanium, gold, indium, lithium, manganese, nickel, niobium, palladium, platinum, rhodium, ruthenium, selenium, silver, tantalum, terbium, thorium, titanium, vanadium, and yttrium.

6.5.16 Applications of Substances and Elements in E-Waste

The list of example applications of the above elements and substances in e-waste are as follows.

Almost all electronics contain lead and tin (as solder) and copper (as wire and PCB tracks), though the use of lead-free solder is now spreading rapidly.

- Lead: solder, CRT monitors (lead in glass), lead-acid batteries
- Tin: solder, coatings on component leads
- Copper: copper wire, printed circuit board tracks, component leads
- Cadmium: light-sensitive resistors, corrosion-resistant alloys for marine and aviation environments
- Aluminum: nearly all electronic goods using more than a few watts of power (heatsinks), electrolytic capacitors.
- Beryllium oxide: filler in some thermal interface materials such as thermal grease used on heatsinks for CPUs and power transistors, magnetrons, X-ray-transparent ceramic windows, heat-transfer fins in vacuum tubes, and gas lasers

- Iron: steel chassis, cases, and fixings
- Silicon: glass, transistors, ICs, printed circuit boards.
- Nickel and cadmium: nickel–cadmium batteries
- Lithium: lithium-ion battery
- Zinc: plating for steel parts
- Gold: connector plating, primarily in computer equipment
- Americium: smoke alarms (radioactive source)
- Germanium: 1950–60s transistorized electronics (bipolar junction transistors)
- Mercury: fluorescent tubes (numerous applications), tilt switches (pinball games, mechanical doorbells, thermostats)
- Sulfur: lead-acid batteries
- Carbon: steel, plastics, resistors; in almost all electronic equipment
- PCBs (prior to ban): in almost all 1930–70s equipment including capacitors, transformers, wiring insulation, paints, inks, and flexible sealants.

Electronic appliances are composed of hundreds of different materials that can be both toxic but also of high value. While bulk materials such as iron, aluminum, plastics, and glass account for over 80% weight, valuable and toxic materials are found in smaller quantities but are still of high importance. The material composition of different appliances is often similar, but the percentage of different components can vary a lot. The average weight and composition of personal computers, televisions, and mobile phones are as given in Table 6.4.

Table 6.4 Average percentage weight and composition of personal computer, television, and mobile phone

Appliances	Average weight	Fe	Non-Fe metal	Glass	Plastic	Electronic components	Others weight
Computer	29.6	53.3	8.4	15	23.3	17.3	0.7
Television	36.2	5.3	5.4	62	22.9	0.9	3.5
Mobile phone	0.08–0.1	8	20	10.6	59.6		1.8

Source: Guidelines for environmentally sound management of e-waste by MoEF and CPCB, March 2008.

Impact type	Quantity used to make your	Reductions through reuse and responsible recycling	
Water	Liters of water 6,950,768	% Reduced 34%	Liters of water reduced 2,393,876
Fossil fuel	Kilograms of fossil fuel 1,277,630	% Reduced 35%	Kilograms of fossil fuel reduced 441,288
Industrial chemicals	Kilograms of chemicals 76,550	% Reduced 51%	Kilograms of chemicals reduced 38,891
Toxic metals	Lethal doses of toxic metal 30,395	% Reduced 62%	Number of lethal doses avoided 18,757

Source: Ramky e-waste recycling facility.

6.6 QUANTITY OF E-WASTE

At the consumer end, disposal of e-waste or used product is a big issue. Among the types of e-waste, computers and peripherals are recycled/reused much more than they are in developed countries. Since the 1990s, affordability of computers was limited to only a socioeconomically advantaged section of the population. Resale and reuse of computers continues to be high as does dependency on assembled machines. No reliable figures are available as yet to quantify the e-waste generation. Increasingly, as computers are becoming more affordable and there is greater access to technology, the turnover of machines could definitely be higher. Apart from the consumer end, another source of more obsolete computers in the market is from the large software industry where use of cutting-edge technology, and greater computing speed and efficiency necessarily increase the rate of obsolescence. In the same way as the standard of living is growing high, dealers are providing monthly payment/installment facilities, and banks are providing loans in a comparatively easy way, affordability of electronic equipment and other household appliances is increasing enormously. As the consumption pattern increases, e-waste generation also increases. The e-waste forms 1% of solid waste on average in developed countries and is expected to grow to 2% by 2010. In developing countries like India, the generation of e-waste ranges in between

Table 6.5 E-waste/WEEE generation in top 10 states

S. No.	Cities	WEEE (tonnes)
1.	Maharashtra	20,270.59
2.	Tamil Nadu	13,486.24
3.	Andhra Pradesh	12,780.33
4.	Uttar Pradesh	10,381.11
5.	West Bengal	10,059.36
6.	Delhi	9729.15
7.	Karnataka	9118.74
8.	Gujarat	8994.33
9.	Madhya Pradesh	7800.62
10.	Punjab	6958.46

Source: E-waste management in India—Consumer Voice, April 2009.

0.05% and 1% of total solid waste generated. In the European Union the average e-waste generated per capita per year is 21.8 kg. In India, the average e-waste generated per capita per year is 0.29 kg.

The quantity of e-waste generated by top 10 states and cities in India are as given in Table 6.5.

S. No.	Cities	Computer, printer, mobile, and TV
1.	Hyderabad	
2.	Bangalore	

Source: Inventonization of e-waste in two cities in Andhra Pradesh and Karnataka—Raiza S. EPTRI.

Northern India is not a leading generator, but it happens to be the leading processing center of e-waste in the country. There are three formal recyclers in the South of India (Chennai, Hyderabad, and Bangalore) and one in Western India.

According to the Manufacturer's Association for Information Technology (MAIT) report, India in 2007 generated 380,000 tonnes of e-waste from discarded computers, televisions, and mobile phones. This was projected to grow to more than 800,000 tonnes by 2012 with a growth rate of 15%. The estimate includes 50,000 tonnes of such e-waste imported from developed countries as charity for reuse, which mostly ends up in informal recycling yards either immediately or once the reused product is discarded. This is a conservative and restricted estimate. Complex, ambiguous definitions of second-hand electronic equipment has made it difficult for the customs department to trace, identify, and stop the illegal in-flow of e-waste (Table 6.6).

Table 6.6 E-waste/WEEE generation in top 10 cities

S. No.	Cities	WEEE (tonnes)
1.	Ahmedabad	3287.5
2.	Bangalore	4648.4
3.	Chennai	4132.2
4.	Delhi	9730.3
5.	Hyderabad	2833.5
6.	Kolkata	4025.3
7.	Mumbai	11,017.1
8.	Nagpur	1768.9
9.	Pune	2584.2
10.	Surat	1836.5

Source: E-waste management in India—Consumer Voice, April 2009.

According to the Environment Protection Training and Research Institute (EPTRI) report, Hyderabad and Bangalore in 2009 generated 3,263.994 MT and 6,743.87 MT of e-waste from discarded computers, printers, televisions, and mobile phones. Around 95,120 kg of e-waste (computers, televisions, and mobile phones) from the household sector is generated in Hyderabad, and in Bangalore it is 121,410 kg. This was projected to grow to more than 107,886 kg in Hyderabad and 1,30,383 kg in Bangalore by 2013.

The following three categories of e-waste/WEEE account for almost 90% of the generation of waste.

1. large household appliances (42%)
2. information and communications technology equipment (33.9%)
3. consumer electronics (13.7%)

6.7 ORGANIZATION AND MANAGEMENT OF E-WASTE

The authorized e-waste recycling facilities in India capture only 3% of total e-waste generated, the rest makes its way to informal recycling yards in major cities like Delhi, Mumbai, Hyderabad, and Bangalore. This is because businesses sell their discarded equipment to informal recyclers for quick money without realizing the hazardous implications it causes to health and environment. E-waste contains over 1000 different substances, many of which are toxic, and creates serious pollution upon disposal. Due to the extreme rates of obsolescence, e-waste produces much higher volumes of waste in comparison to other consumer goods. The increasingly rapid evolution of technology combined with rapid product obsolescence has effectively rendered everything disposable, which causes e-waste to be generated at alarming rates.

The waste thus produced goes into the hands of the informal sector. Over 1 million poor people in India are involved in the manual recycling operations. Most of the people working in this recycling sector are the urban poor with very low literacy levels and hence very little awareness regarding the hazards of e-waste toxins. Women and children are mostly engaged in these activities and they are more vulnerable to the hazards of this waste.

6.8 EVOLUTION OF E-WASTE MANAGEMENT

The Secretariat of the Basal Convention (SBC) has taken a number of initiatives in e-waste management. A pilot project on e-waste management in the Asia and the Pacific Region, in which India has participated, has been supported by SBC. SBC has also facilitated a Mobile Phone Partnership Program (MP3) with a public–private partnership. The MP3 has evolved guidelines for environmentally sound management (ESM) and transboundary movement of mobile phones.

Gesellschaft für Technische Zusammenarbeit (GTZ) and MAIT carried out two studies on e-waste generation, disposal, and recycling of e-waste in Delhi and also in other parts of India. Around 80–85% is recycled by unauthorized recyclers. According to the MAIT-GTZ study, India generates 330,000 MT of e-waste, while 50,000 MT is being imported and the quantity recycled is only 19,000 tonnes. The recycling capacity of authorized recyclers is 60,000 TPA Central Pollution Control Board (CPCB). The recyclables recovery rate is as follows:

- United States: 60%
- Europe: 70%
- United Kingdom: 35%
- China: 25%
- India: 14%

The Department of Information Technology has implemented a project called the Environmental Management in Semiconductor and Printed Circuit Board Industry in India in association with the United Nations Environment Program (UNEP). The electronic production processes were evaluated to explore environmental implications, and promote cleaner production technologies and reduction of hazardous substances in electronic products.

As regards the takeback policy in India, Apple, Microsoft, Panasonic, PCS, Philips, Sharp, Sony, Sony Ericsson, and Toshiba observe the takeback option at their production plant. Samsung claims to have a takeback service but only one collection point for the whole of India; the other nine branded companies do not have a takeback service. Two brands stand out as having the best takeback practice in India: HCL and WIPRO. Other brands that do relatively well are Nokia, Acer, Motorola, and LGE. The details of availability of takeback services, service on ground reality, and accessibility of information on takeback services in India is as given Tables 6.7 and 6.8.

The Central Pollution Control Board (CPCB), with the help of IRGSSA, prepared a status report titled "Management, Handling and Practices of E Waste Recycling in Delhi" in 2004–05. Based on these studies, it was realized that guidelines for the ESM of e-waste is very much essential. As a first step toward ESM, guidelines have been published.

Table 6.7 Availability of takeback services in India

Available in India	Not available in India
Acer, Dell[a], HCL, Hewlett-Packard (HP)[c], Lenovo, LG Electronics[a,b], Motorola, Nokia, WIPRO, Zenith, and Samsung	Apple, Microsoft Panasonic PCS Technology, Philips, Sharp, Sony, Sony Ericsson, and Toshiba

[a]Information regarding takeback in India is only available on global website.
[b]Takeback service is only available for mobile phones.
[c]Takeback service is only available for corporate customers.
Source: An Assessment of E-waste Take back in India, www.designouttoxics.org.

Table 6.8 Takeback service on ground in India

Properly working	Partially working	Not working at all
Acer, HCL, WIPRO	LG Electronics Motorola and Nokia	Dell, Hewlett-Packard (HP), Lenovo and Zenith
Accessibility of information on takeback service in India		
Easily accessible HCL and WIPRO	Partially accessible Acer, Lenovo, Motorola, Nokia	Not accessible Dell, LG Electronics, and Zenith

Source: An Assessment of E-waste Take back in India, www.designouttoxics.org.

The Environment Protection Training and Research Institute (EPTRI) prepared a report titled "Inventorization of E-Waste in Two Cities in Andhra Pradesh and Karnataka (Hyderabad and Bangalore)," sponsored by the World Health Organization (WHO), in 2009. In this report, the status of e-waste mapping for the year 2009 and yearwise projection on generation of e-waste from computers, mobile phones, and televisions till 2013 is presented.

The Hazardous Waste (Management and Handling) Rules, 1989, and amended in 2000 and 2003, have been notified under the Environment (Protection) Act, 1986 which also concerns e-waste.

The Hazardous Waste (Management, Handling and Transboundary Movement) Rules, 2008, has been notified under the Environment (Protection) Act, 1986. This rule also deals with e-waste.

Currently Indian legislation does not have a separate and specific act or rule to tackle e-waste in India. But there exist environmental acts like the Water Act, Air Act, and the Environmental Protection Act that regulate the environmental impact related to any waste management and should be considered when setting up a proper disposal system for any waste. There is a mention of e-waste management under the hazardous waste management, handling, and trans-boundary movement rules under the EP Act, 1986. The Ministry of Environment and Forest, India had then developed a consolidated "E-Waste (Management and Handling) Rules, 2011" that had come into effect from May, 2012 onwards.

6.9 TECHNOLOGIES FOR E-WASTE MANAGEMENT

The technologies involved in e-waste management are dismantling the equipment into various parts, metal frames, power supplies, circuit boards, CRTs, and plastics, which are separated often by hand. The material is shredded and sophisticated expensive equipment separates the various metal and plastic fractions, which then are sold to various smelters and or plastics recyclers. The technology permits the recovery of metals, plastics, and glass in an environmentally friendly way.

Electronic goods are composed of hundreds of different materials, often of high value. Gold, platinum, silver, copper, etc., are valuable materials that recyclers recover from e-waste. A study in 1996 found that more than 50% of the weight of an average desktop computer is in plastics, iron, and aluminum. While precious metals as a percentage of the total weight are relatively small, the concentration of such metals, like gold, is found to be higher in e-waste than found in naturally occurring mineral ore.

Metals, plastics, and glass can be recovered from e-waste such as computers, CPUs, peripherals, servers, printers, fax, copiers, motherboards, printed circuit boards, CDs, floppies, tapes, cartridges, telephones, mobile phones, telecom equipment, televisions, audio and video, dry cells, lithium batteries, fluorescent and CFL lamps, household microwaves, washing machines, industrial, medical, military, space electronics, etc. The technology to date is very simple; no incineration is required. The technology includes manual dismantling, segregation, shredding, crushing, pulverizing and density separation, and precious metal recovery, and involves minimum landfill.

A typical e-waste recycling plant as found in some industrialized countries combines the best of dismantling for component recovery with increased capacity to process large amounts of e-waste in a cost-effective manner. Whole computers and pieces of electronic equipment are shredded into smaller pieces to be more manageable and facilitate the separation of the constituent components. Material is fed into a hopper, which

Figure 6.1 Earth Sense Recycle Private Limited (ESRPL): Machineries, CRTs, and toner-cartridge dismantling table. *Courtesy of Earth Sense.*

travels up a conveyor and is dropped into the mechanical separator, which is followed by a number of screening and granulating machines. The entire recycling machinery is enclosed and employs a dust-collection system (Fig. 6.1).

Every tonne of steel recycled makes the following savings:

- 75% of the energy needed to make steel from virgin material
- 40% of the water required in production
- 1.28 tonnes of solid waste
- Reduction of air emissions by 86%
- Reduction of water pollution by 76%

Every tonne of aluminum recycled makes the following savings:

- 6 tonnes of bauxite
- 4 tonnes of chemical products
- 14 MWh of electricity

It takes 70% less energy to recycle plastics.

It takes 40% less energy to recycle glass.

Every tonne of paper recycled saves 17 fully grown trees.

Precious metals can be recovered, refined, and reused for (gold) plating of watch parts, imitation jewelry, and temple items, etc. A few of the recovery, recycle, and reuse uses of e-waste follow.

Recycling efficiency and recoverable weight of elements in e-waste (from a desktop personal computer, television, and mobile phones) are as given in Table 6.9.

Recycling efficiency and recoverable quantity of elements of televisions are as given in Table 6.10.

Table 6.9 Recycling efficiency and recoverable quantity of elements of personal computer weighing ~32 kg

Material name	Content (% of total weight)	Weight of material in computer (kg)	Use	Location	Recycling efficiency (%)	Recoverable weight of elements
Lead	6.2988	2.016	Metal joining	Funnel glass in CRTs, PWB	5	0.08566368
Aluminum	14.1723	4.5344	Structural, conductivity	Housing, CRT, PWB, connectors	80	3.08389248
Germanium	0.0016	0.000512	Semiconductor	PWBs	0	0
Gallium	0.0013	0.000416	Semiconductor	PWBs	0	0
Iron	20.4712	6.5504	Structural, magnetivity	Housing, CRTs, PWBs	80	4.45453312
Tin	1.0078	0.3232	Metal joining	PWBs, CRTs	70	0.19188512
Copper	6.9287	2.2176	Conductivity	CRTs, PWBs, connectors	90	1.69614576
Barium	0.0315	0.01024		Panel glass in CRTs	0	0
Nickel	0.8503	0.272	Structural, magnetivity	Housing, CRT, PWB	0	0
Zinc	2.2046	0.704	Battery, phosphor emitter	PWB, CRT	60	0.35979072
Tanialum	0.0157	0.00512	Capacitor	Capacitors/PWB, power supply	0	0
Indium	0.0016	0.000512	Transistor, rectifier	PWB	60	0.00026112
Vanadium	0.0002	0.000064	Red phosphor emitter	CRT	0	0
Terbium	0	0	Green phosphor activator, dopant	CRT, PWB	0	0
Beryllium	0.0157	0.00512	Thermal conductivity	PWB, connectors	0	0
Gold	0.0016	0.000512	Connectivity, conductivity	Connectivity, conductivity/PWB, connectors	99	0.000430848
Europium	0.0002	0.000064	Phosphor activator	PWB	0	0
Tritium	0.0157	0.00512	Pigment, alloying agent	Housing	0	0

Ruthenium	0.0016	0.000512	Resistive circuit	PWB	80	0.00034816
Cobalt	0.0157	0.00512	Structural, magnetivity	Housing, CRT, PWB	85	0.00362984
Palladium	0.0003	0.000096	Connectivity, conductivity	PWB, connectors	95	0.00007752
Manganese	0.0315	0.01024	Structural, magnetivity	Housing, CRT, PWB	0	0
Silver	0.0189	0.00608	Conductivity			
			Conductivity/PWB, connectors	98	0.005037984	
Antimony	0.0094	0.003008	Diodes	Housing, PWB, CRT	0	0
Bismuth	0.0063	0.002016	Wetting agent in thick film	PWB	0	0
Chromium	0.0063	0.002016	Decorative, hardener	Housing	0	0
Cadmium	0.0094	0.003008	Battery, blue-green phosphor emitter	Housing, PWB, CRT	0	0
Selenium	0.0016	0.000512	Rectifiers	Rectifiers/PWB	70	0.00030464
Niobium	0.0002	0.000064	Welding	Housing	0	0
Yttrium	0.0002	0.000064	Red phosphor emitter	CRT	0	0
Rhodium	0	Â	Thick film conductor	PWB	50	0
Platinum	0	Â	Thick film conductor	PWB	0	0
Mercury	0.0022	0.000704	Batteries, switches	Housing, PWB	0	0
Arsenic	0.0013	0.000416	Doping agent in transistors	PWB	0	0
Silica	24.8803	7.9616	Glass, solid state devices	CRT, PWB	0	0

Table 6.10 Recoverable quantity of elements in a television

S. No.	Elements	Percentage	ppm	Recoverable weight of elements (kg)
1	Aluminum	1.2		0.4344
2	Copper	3.4		1.2308
3	Lead	0.2		0.0724
4	Zinc	0.3		0.1086
5	Nickel	0.0038		0.013756
6	Iron	12		4.344
7	Plastic	26		9.412
8	Glass	53		19.186
9	Silver		20	0.000724
10	Gold		10	0.000362

Table 6.11 Recoverable quantity of elements in mobile phones

Metal	Weight (g)
Copper	16 (1–3% can be recovered)
Silver	0.35 (3–10% can be recovered)
Gold	0.034 (4–18% can be recovered)
Palladium	0.015 (small quantity)
Platinum	0.00034 (small quantity)

Source: http://www.eoearth.org/article/Cell_phone_recycling.

Recovery and recycling of mobile phones are in the early stages of development. The recoverable quantity of elements are as given in Table 6.11.

6.10 CASE STUDY

E-Parisaraa Pvt Ltd is India's first government-approved e-waste recycling company approved by both the CPCB and Karnataka State Pollution Control Board. It offers an enterprise solution for e-waste recycling technology for the first time in India. The project is located on a 1.5-acre property at Karnataka Industrial Area Development Board, Dobaspet which is about 50 km from Bangalore on the Bangalore–Tumkur Expressway, in the area already earmarked for hazardous waste management by the Government of Karnataka.

E-Parisaraa Pvt Ltd was established in the year 2004 and started operations in September 2005. Since then they have provided a full range of innovative and cost-effective recycling services and solutions that focus on environmentally sound technologies. E-Parisaraa's success has also been based on a resourceful combination of technological innovation, professionalism, and commitment to excellence by all its employees. They consider e-waste not as waste but a resource for recovery in an environmentally friendly manner. Their motto is the 3R's: Reduce, Recycle, and Recover through continuous implementations and innovations in recycling technology. India needs simpler, more cost-effective technologies with maximum resource recovery of metals, plastic, and glass by recycling e-waste in an environmentally friendly way.

Infrastructure: The built-up area of 25,000 sq. ft. of eco-friendly construction uses innovative waste utilization methods of soil cement blocks in place of bricks, thermocol concrete blocks, granite dust in place of river sand, and carbide lime for white washing. Land area is 1.5 acre, closed area 25,000 sq. ft., open area 60,000 sq. ft. The current capacity is 3 tonnes/day; full-scale 10 tonnes/day. The initial investment is Rs. 2.5 Cr, power consumption 66 HP, water consumption 700 lpd, employment potential is 58 (direct), 41 (indirect), and employment (full scale) is 150 (direct), 60 (indirect).

The objective of E-Parisaraa is to create an opportunity to transfer waste into socially and industrially beneficial raw material using simple, low-cost, home-grown, environmentally friendly technology. The project is based on the business model of good material management and decision making to optimize resource recovery from e-waste. E-Parisaraa is also the first among Indian e-waste recyclers to obtain ISO 14001:2004 and OHSAS 18001:2007 certification, by TUV SUD (Fig. 6.2).

They recycle obsolete, end-of-life, and custom debonded computers and peripherals, monitors, telephones, communication equipment, video and audio devices, VCRs, DVDs, TVs, cable equipment, circuit boards, CDs, floppies, printers, fax machines, pagers, and industrial and household electronics. Services include witness crushing, assured destruction, certificate of destruction, downstream material accountability, precious metal recovery, customer-friendly guidance, and consultations (Fig. 6.3).

E-Parisaraa provides e-waste collection and disposal services to its customers who are IT majors, public sector workers, etc. They have written agreements with customers and have been approved by the environmental health and safety audits conducted.

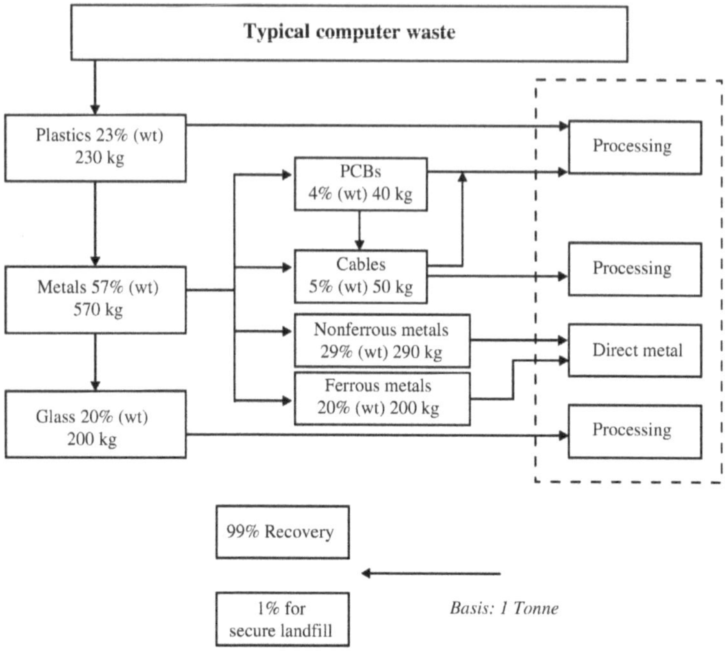

Figure 6.2 Typical computer waste.

The Process/Technology: The process is nonincineration technology, consisting of dismantling, segregation, shredding, crushing, pulverizing, and density separation. They have strong backing of well-equipped laboratory with sample preparation, testing, and analysis; research and development; pollution monitoring; refining; conventional volumetric and gravimetric analysis; and atomic absorption spectrophotometers. The recovered metals, plastics, and glass are sent to authorized recyclers and the hazard part of the e-waste shall be sent to the proposed adjacent TSDF facility (Fig. 6.4).

E-Parisaraa provides all the necessary PPE to its employees. Periodic health monitoring of employees is done. Monitoring of air, water, and noise is done by a third party. They have donated wooden benches, which are made out of waste wooden pallets, to schools. They have conducted seminars on e-waste for students to create awareness among the people, and conducted collection campaigns for employees to create awareness. They provide opportunities to carry out academic projects for both undergraduates and postgraduates under the guidance of experts,

S.No	Material	% composition (by weight)
1	Mild steel	23
2	Stainless steel	8
3	Glass	27
4	Plastics	27
5	Copper	3
6	Aluminium	3
7	Other materials	8
8	Hazardous materials	1

Figure 6.3 Typical computer waste composition (by % weight).

and educate school children. They create job opportunities for rural people in low-tech areas (Fig. 6.5).

If managed safely by recycling, e-waste can be a secondary source of raw material. It will provide an opportunity for revenue generation from recovered materials, natural resource conservation, employment generation, and reduction in pollution. They collect e-waste on a payment basis, provide IPR (intellectual property right) protection, assure data destruction (in front of client representatives, providing a destruction certificate), help in custom debonding by defacing and light destruction, guide customers on legal procedures, and provide logistical support.

E-Parisaraa provides destruction, defacing service for customers for their custom debonding and IPR protection.

Figure 6.4 E-Parisara E-waste component dismantling personnel at work.

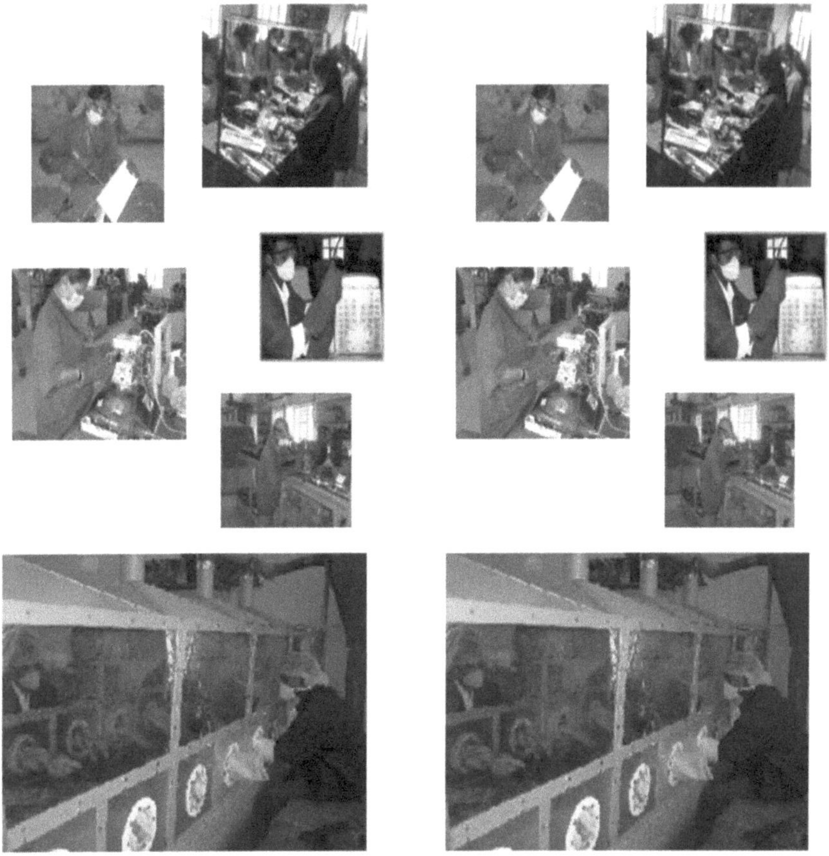

Figure 6.5 E-Parisara E-waste management and handling personnel at work.

6.10.1 Technology

E-Parisaraa has developed its own simple, low-cost machines for recycling of e-waste like printed circuit board shredder, tubelight crusher, CRT-cutting setup, etc.

In its primary stage, equipment like CPUs, monitors, servers, printers, etc., are dismantled and various metal and plastic parts, CBs, cables, different subassemblies, etc., are segregated. Heavier equipment like telecommunication equipment, servers, etc., is handled by men. The lighter equipment like keyboards, mice, hard drives, etc., is handled by women. In the secondary stage, the subassemblies like hard disks, CD drives, floppy drives, etc., are further dismantled and materials are segregated.

Training requirements for the workers are periodically monitored and met. Adequate personal protective aids are provided to all workers. Workers are provided with hand tools, power tools, etc.

They recycle obsolete, end-of-life and custom debonded computers and peripherals, monitors, telephones, communication equipment, video and audio devices, VCRs, DVDs, TVs, cable equipment, circuit boards, CDs, floppies, printers, fax machines, pagers, industrial and household electronics. Services include witness crushing, assured destruction, certificate of destruction, downstream material accountability, precious metal recovery, customer friendly guidance and consultations.

6.11 LEGAL PROVISION

The E-Waste (Management and Handling) Rules 2011, under Environment (Protection) Act, 1986, governs the management of electronic waste. However, the legal provisions change from time to time depending on the minor amendments made in the Rules for handling and management of wastes.

CHAPTER 7

Soil Remediation Technologies

7.1 INTRODUCTION

Releases of hazardous substances have occurred at uncontrolled sites causing soil pollution. Ground water and surface water have been contaminated, drinking water supplies have been lost, and people have been evacuated or in some instances permanently relocated. The general public has become worried and angry about acute and chronic toxic threats to their health and their environment, about loss of natural resources and adverse effects on the value of their homes and property. In many cases state and local governments usually do not have the necessary technological expertise to adequately define either the problem or the solution at the contaminated site. Contaminated soil is one of the environmental problems historically ignored by humans. Lately, its relation to human health and safety, as well as its ecological impact, have been discovered.

7.2 SAMPLING TECHNOLOGY

In many instances, people are unaware that a problem exists until harm has been done. Damage may be in the form of disease to the surrounding population or destruction of the surrounding ecosystem. Monitoring problem areas or potential problem areas can help to limit future damage. Before beginning any sampling program, background research must be conducted to determine proper equipment for both sampling and personal protection, proper sampling methodology and analytical methods, and appropriate health and safety practices to be employed. This is especially important when handling materials that may be hazardous or radioactive. Methods used to obtain data regarding contamination of soil have to take into account factors like the program objective, type of material to be sampled (soil), physical and chemical properties of the contaminant, regulatory requirements and safety, costs, reliability, scale of sample area (small-scale site related to individual persons versus a large-scale site) and short versus long-term sampling requirement.

Solid and Hazardous Waste Management.
DOI: http://dx.doi.org/10.1016/B978-0-12-809734-2.00007-9

Several factors must be accomplished to carry on an adequate sampling practice. The samples must represent the conditions existing at the point taken. A sample must be of sufficient volume and taken frequently enough to permit reproducibility of the testing requisite for the desired objective. The samples must be collected, packed, shipped, and manipulated prior to analysis in a manner that safeguards against change in the particular constituents or properties to be examined. Two portions of the soil that are important to the environmental scientist are the 0−15 cm surface layer and the upper meter. The surface layer (0−15 cm) reflects the deposition of airborne pollutants, especially those recently deposited pollutants. Pollutants that have been deposited by liquid spills or by long-term deposition of water-soluble materials may be found at depths ranging up to several meters. Plumes emanating from hazardous waste dumps or from leaking storage tanks may be found at considerable depths.

Samples can be collected with some form of core sampling or auger device, or they may be collected by use of excavations or trenches. In the latter case, the samples are cut from the soil mass with spades or short punches. Techniques that are utilized should be closely coordinated with the analytical laboratory in order to meet the specific requirements of the analytical methods used.

7.2.1 Surface Soil Sampling

Use thin-walled steel tube that is 15−20 cm long to extract short cores from the soil. The tube is driven into the soil with a wooden mallet, the core and the robe are extracted, and the soil is pushed out of the tube into a stainless steel mixing bowl. Using a seamless steel ring, approximately 15−30 cm in diameter, the ring is driven into the soil to a depth of 15−20 cm. The ring is extracted as a soil-ring unit, and the soil is removed for analysis. Perhaps the most undesirable sample collection device is the shovel or scoop, often used in agriculture.

7.2.2 Shallow Surface Sampling

Sampling pollutants that have moved into the lower soil horizons requires the use of a device that will extract a longer core than can be obtained with the short probes or punches. Three basic methods are used for sampling these deeper soils:

- soil probes or soil augers
- power-driven corers
- trenching.

Samples should be collected at least every 1.5 m or in each distinct stratum. Additional samples should be collected where sand lenses or thin silt and sand lovers appear in the profile.

A laboratory for analyzing the sample should be consulted regarding packaging requirements before the initiation of a sampling program. Samples for organic analysis must be sent to the laboratory urgently. Each sample must be properly documented to ensure timely, correct, and complete analysis for all parameters requested, and, most importantly, to support use of sample data in potential enforcement actions concerning a site. The documentation system provides the means to individually identify, crack, and monitor each sample from the point of collection through final data reporting. To render sample data valid for enforcement uses, individual samples must be traceable continuously from the time of collection until the time of introduction as evidence during litigation. One mechanism utilized is the use of the sample tag. Sampling information recorded on a sample tag includes sample number, station number, date, time, station location, and lab sample number.

The most widely used techniques applied to polluted soils are removal and placement in a more secure landfill environment. Although this simply moves contaminated soil from one place to another, it can be of significant benefit due to improvements in landfill design. Wastes could be stabilized after removal and before or during placement to further reduce mobility after placement. Stabilization might include solidification with concrete or a similar material or direct chemical treatment of certain contaminants. Incineration or thermal treatment of the contaminated soil could be used to eliminate organic contaminants susceptible to destruction or removal by these means. A variety of other processes have been employed to treat contaminated soils once excavated and removed from a site. Included among these are biological degradation in dedicated bioreactors and sophisticated extraction schemes. An alternative to removal options of remediation soil is the use of in situ means that do not require soil removal. These are generally the options of choice if they can be demonstrated effective at reducing the volume, toxicity, or exposure to the wastes.

7.3 TECHNOLOGIES FOR DECONTAMINATION OF SOIL

The principal options to decontaminate soils are removal options for soil remediation and in situ soil remediation processes.

7.3.1 Removal Options for Soil Remediation

These techniques consist of taking the contaminated soil and applying a method to decontaminate it. These techniques are incineration, landfilling, stabilization, solidification, and ex situ bioremediation.

7.3.2 In Situ Soil Remediation Processes

These techniques consist into treat the soil in the same land or place where it is contaminated. The techniques or methods are to pump-and-treat extraction of contaminated ground water, enhancement of treatment processes, vacuum extraction in the unsaturated zone and in situ bioremediation of soils. Pump-and-treat extraction of contaminated ground water is the technique that removes the contaminated ground water or separate contaminated phases via withdrawal wells for aboveground treatment. Enhancement of pump-and-treat processes are methods of remediation of soils. Due to the low solubility of most soil contaminants, large volumes of water are required to remove contaminants present in a separate phase even if it were possible to maintain the water at saturation. Vacuum extraction in the unsaturated zone is a process that is similar conceptually to pump-and-treat of groundwater is soil vacuum extraction (SVE) in the water-unsaturated zone. A vacuum is applied to the unsaturated zone by placing a vacuum pump on a well screened in the unsaturated zone. This pulls vapors through the soil, removing any volatile components that have volatilized in the subsurface. In situ bioremediation of soils is perhaps the most desirable of all treatment processes. In situ biodegradation renders the soil harmless and naturally recycles the contaminants. There are a number of compounds that undergo detoxification by microbial processes at rates that are sufficient to justify natural recovery of contaminated soils.

7.3.3 Site-Remediation Technologies

Successful remediation of sites with contaminated ground water is one of the toughest challenges. In some instances, contamination in ground water will attenuate before it reaches the receptor. In other cases, it is possible to reduce the ground-water contamination concentrations to acceptable levels. The primary focus of risk-based remedial approaches is to mitigate all risks associated with exposure to ground-water contamination by elimination of exposure pathways (i.e., migration in ground water) between contaminant sources and potential receptors. An effective method for achieving this is through the use of horizontal or vertical barrier walls, which represent institutional controls in the form of engineered barriers.

The specific remedial objectives vary from site to site, and the main aim is to minimize or eliminate the hazard to human health and the environment. The cleanup or remediation program must be sufficient to accomplish the objective, i.e., render the site and surrounding environment safe for the intended use. This objective will guide the selection of the appropriate remedial action. In turn, each component of the remedial program will have goals to meet. For some sites, complete removal of the contaminants is the only suitable remedial objective. For others, control of the contaminant migration pathways is adequate and cost effective. In reality no site can be completely cleaned of each molecule of contamination.

One approach to the selection of the appropriate remedial action program is known as risk-based corrective action (RBCA). This approach merges the benefits of risk-based decision making with the other elements of the corrective action process. RBCA is a three-tiered process where each tier represents a different complexity and degree of conservative approach but requires the least tier of effort for the assessment. Advancement to a higher level requires increasing data and analysis, but a more economical corrective action may result as conservative assumptions of the lower level are replaced with site-specific information.

Application of the *Observational Approach* to hazardous waste site remediation requires the recognition of uncertainty and as such suggests: (1) Remedial design based on the most probable site conditions, (2) identification of possible deviations from these conditions, (3) identification of parameters to monitor and confirm conditions and performance of the remedial design, and (4) contingency plans for potential deviation. Recognizing uncertainty and application of the observational approach offers the best opportunity to achieve the desired environmental protection at the lowest cost with the least risk.

Containment is a potential component within remedial systems because this technology may be used for remediation, either alone or with other remedial technologies to accomplish the desired objectives in a cost-effective manner. The technologies are classified as either active remedial system or passive remedial system components.

Active system components require considerable effort and ongoing energy input to operate (e.g., pumping wells).

Passive system components work without much attention, except maintenance (e.g., revegetating a cover soil).

7.3.3.1 Passive Contaminant-Control Systems

The function of passive contaminant-control systems in site remediation is to eliminate exposure pathways (i.e., ground-water migration) and minimize contaminant transport rates. To examine this function, consider the potential pathways for contaminant migration at an uncontrolled hazardous waste site. Precipitation at the site either runs on or off, infiltrates or returns into the atmosphere through evapotranspiration. Precipitation runoff that encounters waste or contaminated soil may transport contaminants or contaminated sediment into the surrounding environment. Precipitation may also infiltrate into and through the waste, generating leachate. Migration of leachate may introduce contaminants into the surface water or ground water. Contaminants present in the leachate migrating to the ground water are the transported through the ground-water environment and may be discharged to surface water in rivers, lakes, ponds, etc.

Passive contaminant-control technologies focus on controlling hydrologic pathways for contaminant migration. This approach is often termed *Containment*. Because of uncertainties associated with the design, construction, and long-term reliability of containment facilities, containment as a sole means of site remediation is not often used for sites with significant levels of contamination. For sites of high waste volume but low risk, containment may be the remedial alternative of choice. Containment is also used at such sites in conjunction with source removal containment and institutional controls.

The selection, design, and construction of containment technologies must recognize each of the exposure and transport processes. Ground-water and containment transport processes include advection, dispersion, adsorption, retardation, and biological and chemical transformation. The selected remedial technology must consider these transport processes to assess its effectiveness in controlling contaminant migration to the surrounding environment.

7.3.3.2 Surface-Water Control Technologies

There is a need to prevent precipitation from transporting contaminants offsite by surface-water runoff or by leachate generation. Covers/caps, surface-water diversions, and sedimentation and erosion control systems are employed for this purpose. Site redevelopment for continued industrial usage could involve covering the site to preclude direct contact, eliminate contaminated runoff, and minimize/eliminate infiltration and leaching to reduce the risk and provide the appropriate level of protection for public health and the environment.

Capillary barriers may limit percolation in semiarid and arid regions. A capillary barrier uses a finer-grained soil overlying a coarse-grained soil. As a result of the particle size difference, downward migration of moisture is limited. In a capillary barrier, moisture is stored in the upper layer of finer-grained material where it is later removed by evapotranspiration.

7.3.3.3 Ground-Water Control

Vertical Barriers: Subsurface vertical barriers are employed to contain contaminants and to redirect ground-water flow. These barriers to horizontal contaminant transport and ground-water flow used in hazardous waste management have developed from conventional engineering applications such as dewatering of excavations and control of ground-water flow through and beneath dams and levees. A number of vertical barrier techniques are available including slurry trench cutoff walls made of soil—bentonite, cement—bentonite, or plastic concrete, and steel sheet piling and grout curtains made of cement-based or chemical grouts.

Passive barrier-wall systems are used in conjunction with conventional pump-and-treat systems to minimize the required pumping rates as well as reduce the rates of chemical migration.

A common function of vertical barrier in a remediation system is to inhibit the flow of clean ground water into the site. With installation of a vertical barrier, clean regional ground water is prevented from entering the pump-and-treat system.

Vertical barriers also provide ground-water control during construction where excavation into the surface is required for direct waste treatment, waste removal, or construction of liner systems. The vertical barrier initially serves to expedite construction.

The principal purpose of an up-gradient vertical barrier wall is to provide a cutoff controlling the influx of clean ground-water flow from up-gradient regions. When used in conjunction with a pumping system located on the down-gradient side, ground-water withdrawal may recover contaminants that have migrated down-gradient of the site. In this way, contaminants that have held site boundaries in total disregard may be captured and returned to the site contaminant treatment program.

The principal purpose of a down-gradient cutoff wall is to utilize site ground water flowing beneath the site to actively flush contaminants from beneath the site. A down-gradient wall is effective in aiding the pump-and-treat system to speed removal of contaminants from the subsurface.

This method of vertical barrier wall deployment was used for remediation at the Rocky Mountain Arsenal site.

Vertical barrier walls are typically embedded or keyed into a material of low permeability beneath the site. In the case of products lighter than water the vertical barrier does not need to fully penetrate the aquifer into the underlying aquifer. However, it is important to consider that ground water has the potential to mound either on the up-gradient side of the barrier or within the barrier, depending on the comparative rates of infiltration and exfiltration. This phenomenon of ground-water mounding will increase the hydraulic head, resulting in enhanced downward migration of ground water. The degree of mounding and subsequent vertical migration of ground water will need to be carefully considered in the barrier design.

Vertical barrier-wall function and configuration are coupled with site and subsurface conditions to select the suitable types of vertical barrier.

7.3.3.4 Types of Vertical Barrier Techniques

- soil–bentonite slurry trench cutoff walls
- cement–bentonite slurry trench cutoff walls
- plastic concrete cutoff walls
- diaphragm (Structural) cutoff walls
- vibrating beam cutoff walls
- deep soil-mixed walls
- composite cutoff walls
- grout curtains

7.3.3.5 Soil–Bentonite Slurry Trench Cutoff Walls

A soil–bentonite slurry trench cutoff wall is also known as a slurry wall. The slurry is composed of approximately 95 percent water and 4–6% bentonite by weight. Bentonite is sodium smectitic clay. The resulting bentonite–water slurry is viscous fluid with a density of 1.03–1.12 g/cc.

Trench collapse is controlled by the resulting hydrostatic force system in which the slurry counteracts the active soil pressures. Because a positive fluid pressure is maintained within the trench, the slurry exhilarates through the walls of the trench, a filter cake is formed and the filtrate enters into the formation. The filter cake is a very thin layer of fully hydrated bentonite, forming an impermeable boundary. The fluid pressure of the slurry opposes the active earth pressure to maintain the trench stability. The excavation material of a vertical cutoff through an embankment is a case along the side of the trench for subsequent use in the preparation of the backfill. The excavation is made in the slurry-filled trench and the excavation sidewalls are essentially vertical. The slurry maintains the trench stability and permits vertical excavations to depths of more than 30 m.

Characteristics to be designed into the soil–bentonite backfill for environmental application are:

1. chemical compatibility
2. low permeability
3. low compressibility
4. moderate strength

In addition to these backfill characteristics, the backfill at the time of construction must flow freely into the trench and displace the bentonite–water slurry used to maintain trench stability. The most important soil parameters that affect these characteristics are grain-size distribution and the water content of the backfill. Soil–bentonite backfills have been successfully fabricated from materials varying from clean sand to highly plastic clay.

7.3.3.6 Cement–Bentonite Slurry Trench Cutoff Walls

Cement–bentonite slurry trench cutoff walls are excavated under a head of slurry to maintain trench stability. The slurry is usually composed of water, cement and bentonite. As the cement hydrates, the combination of the water, cement, bentonite and sand in suspension from excavating results in a hardened material with a consistency of stiff clay or stronger. The cement–bentonite wall is a single-phase process and widely used in the United States.

7.3.3.7 Plastic Concrete Cutoff Walls

Plastic concrete is designed as a mixture of cement, bentonite, water, and aggregate. Cement–bentonite does not include aggregate and has a much higher water content. A plastic concrete trench is usually excavated in panels and plastic concrete is placed in the slurry-filled trench to replace the bentonite–water slurry used for excavation. Plastic concrete may be more contaminant resistant than alternative techniques. Plastic concrete is stronger and less permeable than soil–bentonite.

7.3.3.8 Diaphragm (Structural) Cutoff Wall

In the diaphragm (structural) cutoff wall technique the trench is excavated under a head of bentonite–water slurry to maintain trench stability. However, it is excavated in panels, each 6 m long, and steel reinforcement is placed in the bentonite–water slurry. The slurry is then displaced by high-slump concrete placed by the tremie method of concrete placement.

The diaphragm (structural) cutoff wall technique is widely used in applications requiring significant wall strength, such as where a cutoff wall is to be part of a structural earth-retention system.

7.3.3.9 Vibrated Beam Cutoff Walls

Vibrated beam cutoff walls have been employed to construct vertical barriers for the horizontal control of contaminant migration. A vibratory pile driver specially equipped with injection nozzles at the tip is used to advance a specially modified H-beam into the subsurface. During the driving, grout is injected to lubricate the driven pile. As the beam is withdrawn, a void is left equal to the size of the beam. The void is filled with grout pumped in through the injection nozzles as the beam is withdrawn. Subsequent beam penetrations are overlapped, resulting in a continuous barrier with a typical thickness of 5–8 cm. The slurry may be cement–bentonite or bituminous mixtures. The main advantages of this system are the elimination of the need to excavate potentially contaminated materials, and its success in sandy soils of shallow depths.

7.3.3.10 Deep Soil-Mixed Walls

Deep soil mixing can be employed to construct vertical barriers to horizontal ground-water flows. A special auger mixing shaft was developed that is rotated into the ground while it simultaneously enables the injection of a slurry of bentonite and water or cement, bentonite, and water. Reinforcement can be added to the treated soil columns if additional strength is required. A continuous wall is obtained by overlapping penetrations (0.5–0.9 m wide). The bentonite is added as bentonite–water slurry during mixing; the quantity of bentonite may be limited to 1%. Health and safety risks are minimized by elimination of the need to excavate materials as required for conventional slurry trench cutoff wall techniques.

7.3.3.11 Composite and Active Vertical Barriers

Vertical barrier walls have been constructed in a combination of materials to result in a composite barrier. A geomembrane barrier within a slurry trench cutoff wall is one such composite. The composite cutoff wall properties include lower hydraulic conductivity and increased contaminant resistance.

7.3.3.12 Grout Curtains

Slurry trench cutoff wall technology is limited to soils that can be readily excavated. When it is necessary to install a vertical barrier to a horizontal ground-water flow in material that cannot be readily excavated (e.g., rock), then the most appropriate and commonly used alternative for these conditions, a grout curtain, is installed in rock through the pressure injection of a pumpable material into the pores and fractures. Grout may be made from a variety of materials including cements, silicates, pozzolans, and various chemical formulations. Grouting technology utilized in hazardous waste-containment systems has been developed for conventional geotechnical engineering applications, for beneath dams and levees to reduce seepage.

7.3.3.13 Performance of Vertical Barrier Walls

The long-term performance of containment systems is an important component of the performance since the required design service lives of barrier walls typically range from as little as 10 years to more than 1000 years for radioactive waste storage structures. The management of hazardous wastes requires an assessment of risk. The cause of failure of containment systems such as slurry trench cutoff walls can be classified as construction defects or postconstruction property changes. Construction defects are those anomalies in the barrier wall that can result in greater localized rates of contaminant migration.

In addition to construction defects, long-term property changes in the barrier material must be examined.

Property changes can be result from:

- cycles of freezing and thawing
- cycles of wetting and drying
- desiccation of cutoff wall material
- chemical incompatibility

Cycles of freezing and thawing increase the hydraulic conductivity of the cutoff wall. Cycles of wetting and drying are inevitable with a fluctuating water table and are of particular concern where ground-water fluctuations are evident. Performance of vertical barrier walls is closely associated with the specific site-remediation expectations and objectives.

Key site-specific criteria that can affect both the construction and performance of barrier systems include:

- site-specific land use
- site topography
- equipment accessibility
- site-specific geologic characteristics
- site-specific hydro-geological characteristics
- geotechnical considerations

Construction quality assurance (QA) and quality control (QC) are critical overall factors for the success of a barrier wall project. QA provides a means by which the project stakeholders can verify that the contractor has satisfied the design criteria and minimize the overall risk involved with the performance of the barrier-wall system.

7.3.3.14 Ground-Water Control: Horizontal Barriers

Passive contaminant-control systems include horizontal barriers as well as vertical barriers. Like vertical barriers, horizontal barriers in the subsurface are employed to contain contaminants and to redirect ground-water flow. The subsurface horizontal barriers to vertical contaminant transport and ground-water flow used in hazardous waste management can be categorized by their installation requirements. There are those that can and cannot be constructed in situ. Constructed in situ means that barriers may be constructed beneath an existing waste site or contaminated flume. Most of the horizontal barriers must be constructed prior to the placement of overlying waste or contaminants. These horizontal barriers (or liner systems) were developed from conventional ground-water control applications such as canals, dams, and water-storage impoundments. One function of a horizontal barrier wall is to provide containment for contaminants. By employing a horizontal barrier constructed in situ beneath the site, the rate of contaminant migration is reduced. This is particularly important in sites having dense, nonaqueous-phase liquids (DNAPL). Ground-water pumping systems may not adequately control the downward migration of DNAPL in response to gravitational forces. Passive containment of contaminants by horizontal barriers (i.e., landfill liners) is widely employed as a hazardous waste management-site remediation and can be achieved by excavating wastes and contaminated soil and disposing of them in a landfill.

7.3.3.15 Grouted Liner Systems

Grouting for horizontal barriers has been successfully used to control the influx of ground-water excavations. These applications make it possible to verify the effectiveness of the method. It is expected that grouted horizontal barriers will soon be used for waste-containment applications. Where naturally occurring materials of low permeability, such as clayey soil or intact rock, underlie the site, the naturally low hydraulic conductivity is the preferred bottom containment barrier.

7.3.3.16 Block-Displacement Technique

The block-displacement technique method employs the principles of hydraulic fracturing commonly used to enhance oil recovery. Hydraulic fracturing of the subsurface materials occurs when the fluid pressure exceeds the total stress. The block-displacement technique is actually a special form of grouting; high pressures are used to fracture the soil, and grout is then pumped into the fracture to create the horizontal barrier. This technique has been demonstrated in pilot studies but has not been employed in actual site remediation.

7.3.3.17 Lagoon-Sealing Techniques

In some cases the leakage from the bottom of lagoons can be reduced by installation of a horizontal barrier at the lagoon bottom constructed as the lagoon is still in service. The material used to construct this in situ horizontal barrier is bentonite. Bentonite, a high-swelling clay, can be produced in a granular form. Each grain is similar to particle of sand in size but composed of thousands of colloidal-sized bentonite clay particles. When spread over the surface of a lagoon, the granules sink and form a blanket over the lagoon bottom. In the correct aqueous environment, the bentonite particles hydrate, resulting in significant swelling. The in-place swelling process yields a liner of low permeability. The principal constraint on this system is the chemistry of the lagoon liquids. Liquids that are acidic, organic, or with high-electrolyte concentrations inhibit the hydration of the bentonite and impede the formation of a low-permeability barrier.

7.3.3.18 Active Remediation Systems

Active remediation systems control and contain contaminant migration. Active systems require ongoing energy input and include such systems as pump-and-treat, electrokinetics, in situ biotreatment, and soil washing.

7.3.3.19 Pump-and-Treat Technologies

Hazardous waste management sites have often employed pump-and-treat technologies to remediate contaminated ground water. In this versatile approach, ground water is first extracted by any one of a number of recovery methods including wells, well points and drain-tile collection systems.

The objectives of pump-and-treat systems include:

1. hydraulic containment of contaminated ground water
2. removal of chemical mass from the ground water

Hydraulic containment systems are designed to provide long-term containment of contaminated ground waters at a low cost by optimizing well placement and minimizing pumping rates.

Once the ground water is extracted from the subsurface, any number of water-treatment methods such as air stripping, carbon adsorption, or biological treatment for organics and physical-chemical methods for inorganic may be employed. Thus, pump-and-treat schemes are developed on a site-specific basis for a range of site conditions and a range of contaminant types and concentrations. Although ground-water pump-and-treat systems have shown limited value as a mass removal remedial technology, they are often effective in maintaining hydraulic control and containment of contaminated ground water.

7.3.3.20 Air Sparging

For remediation of ground water contaminated with volatile organics, in situ air sparging may be considered as an alternative to pump-and-treat. The concept of air sparging involves the injection of air into the subsurface (in the zone of saturation) to encourage transfer of volatile organic contaminants from the aqueous phase to the vapor phase. Air sparging is normally conducted in conjunction with soil vapor extraction implemented in the unsaturated zone. Early studies of air sparging revealed that the introduction of air into the subsurface not only resulted in volatilization but also stimulated aerobic biodegradation. When flow rates are designed to stimulate biotransformation, lower flow rates are used than for volatilization and this practice is termed biosparging. The principal benefits of air sparging are the in situ volatilization of dissolved-phase contaminants, desorption and volatilization of contaminants adsorbed to the soil matrix, and the enhanced aerobic biodegradation.

For the remediation of ground water contaminated with volatile organics, in-well stripping may be considered as an alternative to pump-and-treat.

Air stripping is simply the transfer of contaminants from one medium, i.e., water, to another, i.e., air. The reduction in the contaminant concentrations in the ground water provides the necessary protection to public health while the increase of contaminants in the atmosphere may be inconsequential. In some cases, the air may require treatment by techniques such as activated carbon. Air stripping is effective only for volatile and semivolatile compounds. Stripping towers may be inclined to foul because of iron, algae, bacteria, fungi, and fine particulate matter in the water.

7.3.3.21 Electrokinetic Phenomena

Application of electrical fields in fine-grained soils (silt and clay) results in electrokinetic phenomena that influence transport of water, charge, and mass. Electrokinetic is defined as the physicochemical transport of charge, action of charged particles, and effects applied potential on formation and fluid transport in porous media.

Electrokinetic phenomena in soils are developed as discrete clay particles have a negative charge that influences and controls particle movement. The soil particle surface charge can be developed in different ways including the presence of broken bonds and due to isomorphous substitution. Electrokinetic phenomena identified in soils include electroosmosis, electrophoresis, streaming potential, and sedimentation potential. Electroosmosis is defined as fluid movement with respect to a solid wall as a result of an applied electric potential gradient. Electrophoresis is the movement of charged particles suspended in a liquid due to the application of an electrical potential gradient. Stream potential is the reverse of electroosmosis. Sedimentation potential is an electric potential generated by movement of particles suspended in a liquid. Electrokinetic phenomena that affect the electrokinetic remediation are electroosmosis and electrophoresis.

7.3.3.22 Drain-Tile Collection Systems

A drain-tile collection system is constructed by excavating a continuous trench in the subsurface and installing collection piping and filter media. By far Love Canal is the best known use of a drain-tile collection system as a component of site remediation. The continuous nature of the drain-tile collection system intercepts a variety of subsurface features that otherwise allow water to circumnavigate its way around a pumping well, including sand seams, root holes, buried conduits, and prior surface-drainage ways that have been filled in. Thus drain-tile collection systems are more effective in the presence of subsurface heterogeneity and anisotropy. In the case of a floating contaminant, the drain-tile collection system may require only a shallow interception of the ground-water level. For very deep trenches, worker safety is the main problem.

7.3.3.23 Biopolymer Extraction/Interception Trenches

The biopolymer slurry trench is a construction technique whereby a continuous subsurface drain can be installed relatively rapidly, safely, and cost effectively. The technique builds on long-known and employed slurry trench methods to install a subsurface drain.

In the biopolymer drain technique, the slurry is formed by a mixture of biodegradable materials, additives and water. In the biopolymer technique, the trench is backfilled with porous drainage media. Additives to the trench result in the degradation of the slurry to water and natural carbohydrates.

The ground water may be extracted from the trench through well casings lowered vertically in the trench. Gravel is around the well to maintain its vertical position while the remainder of the trench gravel backfill is pushed into the trench. Horizontal drain pipes can be installed along the bottom of the trench by using flexible pipe and a pipe-laying machine that follows along after the excavation.

7.3.3.24 Soil Vapor Extraction

Soil vapor extraction is a remedial option utilized to remove volatile organic compounds from stockpiled excavated soils. The soil vapor extraction process consists of passing an air stream through the soil, thereby transferring the contaminants from the soil (or water) matrix to the air stream.

Modifications of the process are distinguished by the location of the treatment system and the method for developing airflow. Soil vapor extraction systems are implemented by installing vapor extraction wells or perforated piping in the zone of contamination and applying a vacuum to induce the movement of soil gases. Soil vapor extraction systems typically include knockout drums to remove moisture from the soil gases, followed by vapor-phase treatment prior to discharge to the atmosphere.

Soil vapor extraction systems may be enhanced through the addition of alternative options/enhancements including:

- installation of ground water extraction pumps into the vapor extraction wells to either lower the ground-water table and concurrently remove contaminated ground water;

- placing an impermeable barrier over the surface to minimize short circuiting of airflow from the surface and thereby increase the radius of influence;
- installing air recharge wells around the zone of contamination to enhance movement of soil gases through contaminated soils;
- providing a compressor to force clean air into the recharge wells, enhancing soil gas movement;
- installing wells into the zone of contaminated ground water and blowing air through the ground water (air sparging). The induced airflow enhances the volatilization of the contaminants.

Three variables control the performance of an in situ soil vapor extraction system and must be considered during system design:

1. the extraction well spacing
2. the airflow rate induced in the subsurface remedial zone
3. the subsurface pressure

Additional variables relating to the specific contaminant that must be evaluated for the design include:

1. pressure gradients
2. identity and concentration of volatile organic compounds in the soil
3. extracted air volatile organic compounds concentration
4. extracted air temperature
5. extracted air moisture
6. power usage

7.3.3.25 Permeable Reactive Treatment Walls

Recent innovations in remedial technologies have resulted in the widespread application of permeable reactive treatment wall technologies and also known as permeable reactive barriers for ground-water remediation.

Treatment walls are permeable semipermanent or replaceable units that are installed across the flow path of a contaminant plume. Contaminants present in the ground water are removed by physical, chemical, and biological processes that include fixation, degradation, oxidation, reduction, sorption, and precipitation. Each of these reactions depends on a number of parameters such as pH, oxidation/reduction potential, chemical concentration, and kinetics.

Treatment walls are not designed to contain ground-water flow, but rather to contain the chemical mass present in ground water while allowing uncontaminated (treated) ground-water flow to pass through the wall.

Treatment technologies are also passive remedial technologies and therefore do not require a continuous input of energy to run pumps or treatment systems. Operation and maintenance costs are thus significantly reduced. However, periodic replacement or rejuvenation of the reaction medium may be required after the capacity of the treatment is exhausted or clogged by precipitation. Treatment wall designs can be classified as one of two types:

- in situ reaction curtains
- funnel-and-gate systems.

In situ reaction curtains involve the construction of a trench on the down-gradient portion of a ground-water plume, which extends the entire width of the plume. Reactive materials that comprise the treatment wall are installed in the trench and contaminated ground water is remediated as it passes through the treatment wall. The permeability of reaction curtains needs to be higher than or equal to the permeability of the surrounding geologic material to have ground-water flow converge on the curtain. In the event that hydraulic conductivity of the treatment wall is less than the surrounding aquifer material, the ground water will tend to migrate around the treatment wall.

The combination of cutoff walls and reactive curtains is known as the funnel-and-gate system. By combining the low-permeability cutoff walls (funnel) with the treatment wall (gate), the treatment wall is not required to extend the entire width of the contaminant plume, since the funnel acts to focus ground-water flow into the gate and prevent lateral movement of contaminants around the gate. The primary advantage of this method includes the control of ground-water flow and a reduced length of treatment wall.

7.3.3.26 Degradation Treatment Walls

Degradation barriers are designed to break down or degrade contaminants in ground water into harmless products by employing either enhanced biological degradation or chemical destruction mechanisms. Biologic degradation walls are often permeable treatment walls in which nutrients are injected to enhance the biological activity within and down-gradient of the wall.

CHAPTER 8

Waste Minimization

8.1 INTRODUCTION TO WASTE MINIMIZATION

Waste minimization or reduction at source is the most desirable activity, because it does not incur expenditure for waste handling, recycling, and disposal of waste that is never created and delivered to the waste management system. However, it is an unfamiliar activity as it has not been included in earlier waste management systems.

To reduce the amount of waste generated at the source, the most practical and promising methods appear to be the adoption of industry standards for product manufacturing and packaging that use less material, passing of laws that minimize the use of virgin materials in consumer products and levying of cess/fees for waste management services that penalize generators in case of increase in waste quantities.

Waste minimization lowers the costs of production, end-of-pipe treatment, healthcare and cleaning up the environment. It lowers risks to workers, communities, consumers of products, and future generations. Waste minimization prerequisites are willingness, commitment, open mind, teamwork, and structured methodology. The material and energy balances are important component of waste minimization as they are not only used to identify the inputs and outputs of mass and energy but their economic significance is related to costs, such as cost of raw material in waste, final product in waste, energy losses, handling waste, transporting waste, solid waste disposal, and pollution charges and penalties.

Waste minimization reduces costs, increases process efficiency and productivity, maintains or increases competitiveness, decreases exposure to long-term liability, reduces present and future regulatory burdens, improves workplace safety and environmental quality, and maintains or improves institutional image. The principles of reduce, reuse, recycle, and recover are basics for waste minimization.

Reduce: Avoid unnecessary waste generation in the first place, eliminate unnecessary consumption, refine industrial and commercial processes to reduce waste, avoid unnecessary packaging, substitute reusable for disposal and buy durable, long-lasting items.

Solid and Hazardous Waste Management.
DOI: http://dx.doi.org/10.1016/B978-0-12-809734-2.00008-0

Reuse: Use objects, devices, or substances again, refillable containers, durables instead of disposables and reusable packaging.

Recycle: Use waste materials in place of virgin materials to create a new product. There are many recycling variants; to be an appropriate strategy, the net environmental impacts must be lower than the impacts of using virgin materials.

Recover: Extracting energy or material resources from waste, energy recovering, refuse-derived fuel facilities, materials recovery like gold, silver, etc. from e-waste. The recovery is perceived to be contrary to reduce, reuse, and recycle and in the absence of these recovery is an option but care should be taken that there is not any production of toxic emissions.

8.2 MUNICIPAL SOLID WASTE MINIMIZATION

Sorting at source, reuse and recycling at source, and processing at source (e.g., yard composting) help in waste minimization (Fig. 8.1). Modification in product-packaging standards and use of recyclable materials can result in reduction of waste material.

One of the waste management strategies used in some developed countries is to charge a variable rate per can (or tonne) of waste, which gives generators a financial incentive to reduce the amount of waste set out for collection. Issues related to the use of variable rates include the ability to generate the revenues required to pay the costs of facilities, the administration of a complex monitoring and reporting network for service, and the extent to which wastes are being put in another place by the generator and not reduced at source.

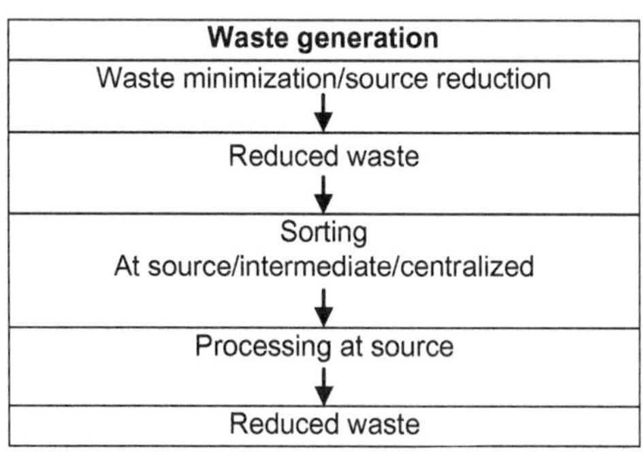

Figure 8.1 Waste generation.

8.3 INDUSTRIAL WASTE MINIMIZATION

Minimization of use of virgin raw materials by the manufacturing industry promotes substitution by recycled materials. Waste minimization is a new and creative way of thinking about products and processes that make them. It is achieved by the continuous application of strategies to minimize the generation of wastes and emissions. Waste minimization needs to include cost reduction, chemical and auxiliaries conservation, water conservaton, and market requirement (ISO 14000). The benefits of waste minimization are higher profits, improved quality, reduced treatment costs, better shop-floor environment, and better image. It lowers the costs of production, end-of-pipe treatment, and cleanup of the environment. It lowers the risks to the workers, the community, consumers of products, and future generations. Waste minimization makes good business and environmental sense. It does not necessarily require large capital investment. It forms a sound basis for any further pollution-control measures and has been demonstrated in many parts of the world including India.

The prerequisites for waste minimization are willingness, commitment, open mind, teamwork, and structured methodology. Waste-minimization assessment is a systematic planned procedure with the objective of identifying ways to eliminate or reduce the generation of waste and emissions. The waste-minimization assessment consists of six steps:

1. getting started
2. analyzing process steps
3. generating waste-minimization options
4. selecting waste-minimization solutions
5. implementing waste-minimization solutions
6. sustaining waste minimization.

Waste-minimization techniques include good housekeeping, input material change, better process control, equipment modification, technology change, recycling/reuse, recovery, byproduct manufacture, and product modification. Various waste-minimization techniques are presented in Fig. 8.2.

Material and energy balances are very important not only to identify the inputs and outputs of mass and energy, but because their economic significance is related to the cost of raw material in waste, final product in waste, energy losses, handling waste, transporting waste, solid waste disposal, and pollution charges and penalties. Various waste management practices include:

- good housekeeping that takes appropriate managerial and operational actions to prevent leaks and spills;

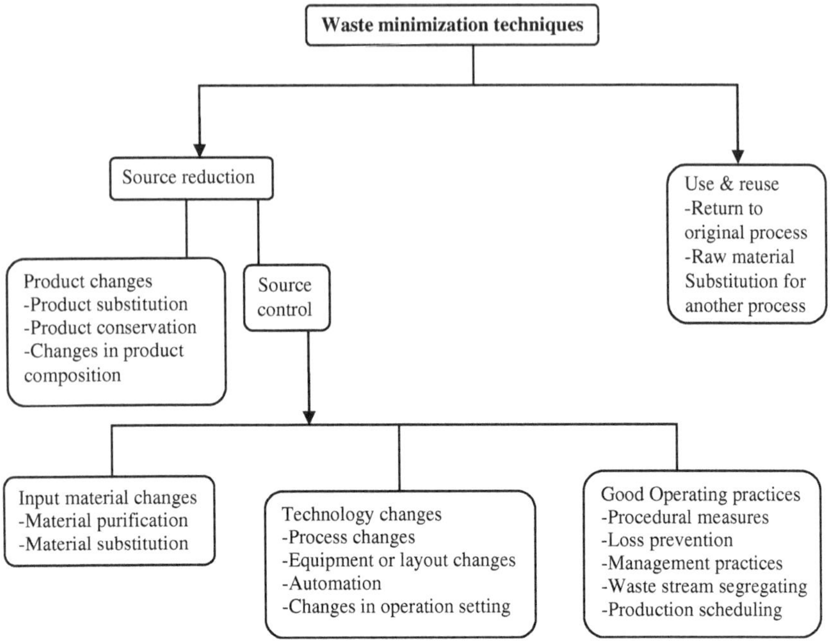

Figure 8.2 Waste-minimization techniques.

- substituting input materials by less toxic or by renewable materials or by materials that have a longer service life time in production;
- looking for better process control by modifying operational procedures, equipment modification, and process record keeping in order to run the processes more efficiently and at lower waste and emission generation;
- reuse of the waste materials in the same process for another useful application within the company;
- production of a useful byproduct by transforming waste into a useful byproduct, to be sold as input for companies in different business sectors;
- modifying the product characteristics in order to minimize the environmental impacts of the product during or after its use (disposal).

The other terminologies of waste minimization include:

- cleaner production
- pollution prevention
- pollution preventon pays
- green productivity.

Waste-minimization options can be broadly divided into three levels, as described below. Industries often employ a number or combination of these approaches simultaneously to resolve a particular waste management problem.

Level	Investment	Potential of waste reduction
1. Good housekeeping and direct reuse, awareness, housekeeping, reexamination of operating procedures, direct reuse wherever possible, establishment of monitoring information system	Low	Moderate
2. Resource recovery from waste stream; characterization and segregation of waste streams; recycle, reuse, recovery at the manufacturing process as well as the waste treatment facility	Moderate to high	Moderate to high
3. Manufacturing changes, waste minimization via cleaner technologies, modifications, replacement of raw material equipment as well as the process	High	High

8.3.1 Level I: Good Housekeeping

Good housekeeping aims to operate machinery and manufacturing systems in the most efficient manner. As such, it is a basic task of management. For example, the proper operation and regular maintenance of equipment can often substantially reduce leakage and overuse of materials. Improvements in housekeeping practices, which can often reduce waste by between a quarter and a third, usually do not require large capital expenditure.

In many industries, it has been noticed that a quarter of a company's environment costs come from losses in containment (i.e., during storage, transfer, and handling of materials) and suboptimal operation of plant and textiles; significant water conservation has been achieved through relatively simple housekeeping practices. The volume of waste water can be substantially reduced by such obvious practices as shutting off of water supply to equipment not in use, installing automatic shutoff valves in houses supplying only the optimum water to the machine. Major targets of water reduction or reuse are the cooling water systems including boiler operation. This must be tightened up on a rational basis.

Water consumption and waste generation can also be reduced through recycling programs. Textile companies can lower water use by installing countercurrent washing: the least contaminated from the final wash is reused for the next-to last wash and so on, until the water reaches the first wash stage where it is discharged from the system and treated. Furthermore, the use of hot rather than cold water in some cases has halved consumption in the textile industry.

A unit manufacturing pharmaceutical formulations is utilizing its waste water for floor- or vessel-cleaning purposes. In yet another unit manufacturing synthetic textiles, the effluent is being recycled for the humidification plant.

Good housekeeping requires attention to detail and monitoring of raw material flows and impacts. Many companies still have no idea of how much or what type of wastes and pollution they produce. Waste minimization begins with accurate measurement, identification, and then separation of wastes. Improvements in information technology have also made environmental monitoring more affordable. Some companies have obtained clear benefits from introducing sophisticated waste-measuring and tracking systems.

In a multinational pharmaceutical company in Europe, the very quality that made its dye product so desirable, i.e., its high color intensity and stability against a range of chemical, physical and biological effects, would have made untreated effluent from the production process impervious to normal biological treatment. Discharging without treatment is out of question and to install a treatment plant that could remove the color from the effluent would cost more the standardization of the plant itself. It was found after studies that the cleaning process required between each new dyestuff was the source of highly colored water. Since the unit produces a variety of different color dyes, the plant has to change over an average twice a week. This involves a complete clean out of the production unit to prevent any cross-contamination of dyes.

Instead of discharging this water, the firm decided to collect and store it, and to recycle it the next time the dyestuff was produced. To minimize the storage of this waste water, the company also designed the process to minimize water use. Though the company has to keep about 300 MT of water in storage, which takes up one-third of the plant space, the new closed-loop production system has enabled the company to recover 3% more dye from the water. In only 2 years, this covered the cost of necessary investments. Chemical oxygen demand and Biochemical Oxygen Demand (BOD) loads were reduced to almost 75% in spite of the 25% increase in production.

It is a well-known fact that water is the main carrying medium in the paper manufacturing process, and the rate of water consumption per tonne of paper is generally very high. The pollution control board has enforced very stringent norms, particularly on the rate of water consumption and rate of effluent discharge from a large integrated pulp-and-paper unit, i.e., 175 m^3/tonne of paper.

A pulp-and-paper unit in Andhra Pradesh made special efforts to conserve water by adopting technologies like effective recycling of back waters, arresting leakages, minimizing spillage, creating awareness among workmen, etc., so as to bring down effluent discharge to the lowest possible level. Further, certain process modifications such as chlorine dioxide bleaching in place of conventional bleaching was adopted resulting in considerable reduction of water consumption. They adopted various technologies for water conservation. Detailed and comprehensive studies were conducted on various unit processes for upgrading and effective disposal of solid waste, to mitigate odor, and to further minimize water consumption. The rate of effluent discharge, which in 2–3 years averaged around 300 cubic meters per tonne of paper produced, has been reduced to 158 cubic meters per tonne of paper produced. There are a number of similar instances in Indian industries where good housekeeping, rationalization of operating conditions, and direct reuse have given substantial benefits. A few illustrating examples, reflecting a range of different situations, are given below.

8.3.1.1 Direct Recycling From Water-Ring Vacuum Pumps in a Pharmaceutical Unit

It is a normal practice in the chemical industry to use water-ring vacuum pumps for the creation of a vacuum. Normally, the water from these pumps is drained to a treatment plant as it is contaminated with organics. In a pharmaceutical company in Mumbai, about 150 m^3/day of such water is collected in pumps in sump and recycled in the system (i.e., used again and again for the same water-ring vacuum pumps). In order to avoid corrosion, smell, etc., it was found by experience that the level of COD of about 500 mg/L works out as a reasonably good cutoff point for draining the waste water. It was found that initial COD was about 30 mg/L and it takes about 20 days to reach the value of 500 mg/L COD. Thus for 20 days at the rate of 150 m^3/day, total water saved due to recycling was 3000 m^3 or 37,620 m^3/annum (22 days/month operation). This is equivalent to a sum of about Rs. 450,000/annum on account of reduced water consumption. A capital investment of Rs. 20,000 was made for construction of the sump.

8.3.1.2 Water Reuse and Recycling in the Fertilizer Unit

In a cooperative fertilizer unit in Surat, the dividends of water reuse and recycling have been indeed impressive. This has been achieved through in-house development as well as help from external consultants from abroad. The cost of these measures worked out to Rs. 20 crores, with a payback in less than 8 years.

Reuse of treated municipal waste water for industrial and other purposes has been practiced in some industries in India since 1967, especially in areas like Mumbai, Gujarat, and Chennai where water scarcity is very high, there is seldom any alternative other than going for reuse with treatment. A multinational chemical company in Mumbai installed a similar sewage-treatment and renovation plant to treat 5000 m^3/day of raw sewage comprising domestic as well as industrial effluent. About 90–93% of the raw sewage inflow was converted into reusable water, which was used as cooling water.

Recycling of treated effluent in process manufacturing is, however, an area requiring greater control and studies. The following case study of the petrochemical unit emphasizes this point.

8.3.1.3 Reuse of Treated Waste Water in Process Manufacturing in a Petrochemical Unit

A petrochemical unit near Chennai practices recycling of treated waste water to the propylene oxide reactor. In the propylene oxide reactor, raw water is used to dissolve chlorine into hypochlorous acid. It is then reacted with vapor propylene from chlorohydrins solution. The quantity of water requirement is as high as 65 m^3/hour for the full load condition. The treated waste water contains high salinity. As a part of waste minimization, the treated waste water led to a foaming problem inside the reactor, thus causing a lot of pressure drop between the reactor and the downstream equipment. Further, the salinity or effluent generated after reuse was observed to be very high, thereby making biological treatment difficult. In view of the above problems, the recycling of treated waste water is restricted to only 20 m^3/hour, i.e., 30% of the requirement.

Another low-quality requirement of water use was noticed at the stage of milk or lime preparation. A portion of treated water was hence diverted for this application. However, it was noticed that the required concentration of milk of lime is not achieved on reuse. Hence, usage of treated waste water was once again limited to 8–10 m^3/hour.

8.3.1.4 Level II: Resource Recovery From the Waste Stream

What separates a raw material from waste is its economic usefulness. The case studies that follow demonstrate how Indian companies have actually used research and imagination to turn "wastes" into "resources" by practicing resource recovery.

A unit manufacturing vanaspati and chemicals recovers free-floating oil and fatty acids and utilizes them for the manufacture of soap cake. In the same unit, the excess biosludge is dried and used as fertilizer.

A unit manufacturing complex chemicals has seven streams flowing out from the manufacturing process as effluent. Out of these, four streams contain zinc. These streams are treated with NaOH to obtain zinc hydroxide as a byproduct, which is used in the manufacturing plant. Around 30 tonnes/month of zinc hydroxide are recovered from the effluent. One stream of the effluent coming from the zinc treatment is used as raw water in other plant. The quantity used is around 40,000 liters/day. Another stream of 25% sodium chloride solution is reused; the amount reused is 10,000 liters/day.

There are basically two options for implementation of cleaner technologies:

1. Attempt to enhance the existing manufacturing efficiency to the extent possible via reuse and recovery.
2. Modify the manufacturing process either in terms of new operating parameters, equipment, modifications, or new pathways and raw materials.

The first option has been popular, as is reflected in the experience of industrialized developed countries. Its wider application requires the development and establishment of cross-sectoral or "hub" technologies. Examples of such technologies are membrane filtration, electrolysis, adsorption, and ozonation, etc. Currently these technologies are not widely used on a commercial scale in Indian industries.

The second option for examining manufacturing process changes is a less preferred alternative. This attitude has been mainly due to lack of know-how, high cost of capital, and inertia to change. These technologies may be termed sector-specific manufacturing technologies. Examples of these are transfer printing ink in textiles, chrome-free tanning in the leather industry, continuous and low-pollution bleaching in the pulp and paper industry, etc.

There are several such examples of how the apparent disadvantage of waste generation can be converted into an economic advantage by seeking innovative solutions. These examples also demonstrate how increasingly stringent pollution control regulation has helped trigger innovative byproduct recovery or recycling systems in Indian industries. A few of these examples are cited below to help the reader to better appreciate the technoeconomic potential of this option.

8.3.1.5 Byproduct Recovery From Hydrogen Sulfide in a Pesticide Unit

A pesticide unit faced the problem of handling the sulfide that was generated during the manufacture of organophosphorus insecticides. Hydrogen sulfide gas is generated in the reaction of phosphorus pentasulfide and alcohol. Earlier, this gas was flared out in a flare stack using Liquefied Petroleum Gas (LPG) as an auxiliary fuel, as its release otherwise would cause severe odor nuisance. As an alternative strategy, the gas is now absorbed in a caustic soda solution and a byproduct, NaSH, is produced that has great demand as a reducing agent in dye manufacturing. Around 75 MT of hydrogen sulfide is now scrubbed to give the byproduct, NaSH. The economics of this modification were favorable; for a capital investment of Rs. 10 lakhs, payback was achieved in only 10 months.

8.3.1.6 Byproduct Recovery From Sulfur Dioxide in a Dye and Dye Intermediate Unit

A second example of a dye and dye intermediate company can be cited. This company manufactures 1300 tonnes/year of azo- and nonbenzidine dyes and dye intermediates. The raw material is naphthalene, which is subjected to sulfonation and nitration, followed by reduction, isolation, and filtration and dyeing. In the isolation step sulfuric acid is used to get different dyes and intermediates and in this step sulfur dioxide is evolved.

Prior to recovery of implementation scheme sulfur dioxide was being scrubbed by sodium hydroxide resulting in the formation of sodium sulfite. The sodium sulfite stream is treated in the effluent treatment plant. As the scale of operation increased, the economics of sulfur dioxide recovery became attractive. Ammonium sulfite is required for the production of some of the dye intermediates; utilization of the recovered sulfur dioxide for production of the ammonium sulfite was an additional motivation. At a capital investment of Rs. 20 lakhs, which was subsidized by financing institutions, a payback period of around 36 months was obtained.

8.3.1.7 Level III: Manufacturing Changes: Waste Minimization via Cleaner Technologies

It is worth restating in detail the difference between waste minimization and cleaner technology vis-a-vis the conventional approach to waste treatment in contrast to end-of-pipe technologies. The implementation of cleaner technology requires large-scale changes in existing processes or their total replacement. In addition, cleaner technologies are highly process-specific, even when the basic technologies already exist. Considerable development is often required to generate a solution that is usable by an individual company.

Considerable pollution is caused through outdated and inefficient technologies and poor process control. Look for change in the production process. Change in raw material by using different types or physical forms of catalysts, use water-based coatings instead of VOC-based coatings, pure oxygen instead of air for oxidation reactions, terpene or carbon dioxide instead of chlorinated or flammable solvents, plastic blasting media or dry ice pellets instead of sand blasting, dry developers instead of wet developers for nondestructive testing and hot-air drying instead of solvent drying for components. Optimize the relative location of unit operations within a process, investigate consolidation of unit operations where feasible, optimize reactor design alternatives to the continuously stirred tank reactor, investigate a separate reactor for processing recycling and waste streams, different ways of adding reactants (e.g., slurries vs solid powders), changing the order of adding reaction raw material, chemicals synthesis methods based on renewable resources rather than petrochemical feedstocks and conversion of batch operations to continuous operations. Operational changes can be brought, e.g., a computerized material inventory system should be introduced; for example, reduction of emissions of cyclohexane solvent from storage and loading and unloading operations in a synthetic organic chemical manufacturing industry (SOCMI).

Table 8.1 Savings from products manufactured using recycled materials

	Paper (%)	Glass (%)	Steel (%)	Aluminum (%)
Energy	23–70	4–22	47–74	92–97
Air pollution	74	20	86	95
Water pollution	35	–	76	97
Mining wastes	–	80	97	–
Water use	58	50	40	–

Source: Patrick Walsh and Phil O'Leary. (1988). Recycling Offers Benefits, Opportunities And Challenges. Waste Age. http://infohouse.p2ric.org/ref/10/09714.pdf.

Table 8.2 Waste exchange options

Waste generator industries	Waste	Waste user	usage
Chemical	Gypsum powder	Cement	Cement
Electronics	Organic solvent	Chemical	Paint solvent
Food	Noodle chips	Livestock	Animal feed
Plastic	Spent plastic	Plastic	Artificial wood
Thermal power plant	Fly ash	Cement	Cement

A major way to reduce waste in the manufacture of complex organic substances is to reduce the number of steps required from raw materials to the final product (Table 8.1).

Waste exchange information makes industrial wastes available to potential users of those materials and enables industrial wastes to be transferred from one company to another where it may be used as an input material. Waste generator industries, users, and usage is as given in Table 8.2.

CHAPTER 9

Environment Impact Assessment

9.1 INTRODUCTION TO ENVIRONMENTAL IMPACT ASSESSMENT

Every anthropogenic activity has some impact on the environment. It is necessary to take up the activities for food, security, and other needs. There is a need to harmonize developmental activities with environmental concerns. Environmental impact assessment (EIA) is one of the tools available to planners to achieve this goal. It is desirable to ensure that the development options under consideration are sustainable. In doing so, environmental consequences must be characterized early in the project cycle and accounted for in the project design. The objective of EIA is to foresee the potential environmental problems that would arise out of a proposed development and address them in the project plan and designing stage. It integrates the environmental concerns in the developmental activities right at the time of initiating for preparing the feasibility report, which will enable the integration of environmental concerns and mitigation measures in project development. It can often prevent future liabilities or expensive alterations in project design. It is a policy and management tool for both planning and decision making. It assists in identifying, predicting, and evaluating the foreseeable environmental consequence of proposed development projects.

The EIA is defined as:

- any alteration of environmental conditions or creation of a new set of environmental conditions, adverse or beneficial, caused or induced by the action or set of actions under consideration;
- a set of procedures which permit an understanding of the likely consequences of society's economic growth activities on the environment;
- an activity designated to identify and predict the impacts on society's health and well being of legislative proposals, policies, programs, projects and operational procedures and interpret and communicate information about the impacts;

Solid and Hazardous Waste Management.
DOI: http://dx.doi.org/10.1016/B978-0-12-809734-2.00009-2

- an attempt to evaluate the consequences of a proposed action on each of the descriptors in the environmental inventory;
- the administrative process by which the environmental impacts of a project is determined;
- a policy and management tool for planning and decision making which assists to identify, predict, and evaluate the foreseeable environmental consequences of proposed developmental projects, plans, and policies. The outcome of an EIA study assists the decision maker and the general public to determine whether a project should be implemented or not. EIA does not make decisions, but it is essential for those who do;
- the documentation of an environmental analysis, which includes identification, interpretation, prediction, and mitigation of foreseeable impacts caused by a proposed action or project.

Under Environment (Protection) Act, 1986, EIA Notification was issued in 1994 and its subsequent amendments issued in the years 1994, 1997, 2000, and 2006, making EIA statutory for different types of 30 categories of activities/industries based on their significant probability of hazard/accident. EIA is one of the tools available to planners to ensure that the development options under consideration are sustainable. The objectives of EIA are:

- to ensure that environmental aspects are addressed and potential problems are foreseen at the appropriate stage of project design;
- to integrate the environmental concerns in the developmental activities right at the time of initiating for preparing the feasibility report;
- to help the decision makers to protect, conserve, and manage the environment according to the principles of sustainable development, thereby achieving or maintaining human well being, and a healthy and sound environment.

The environment-sensitive areas are religious and historic places, archeological monuments/sites, scenic areas, hill resorts/mountains, beach resorts, coastal areas rich in corals, mangroves, breeding grounds of specific species, estuaries, biosphere reserves, national parks, wildlife sanctuaries, lakes, swamps, seismic zones, tribal settlements, areas of scientific and geological interests, defense installations, especially those of security importance and sensitive to pollution, border areas (international), airports, tiger reserves/elephant reserves/turtle nesting grounds, habitats for migratory birds, reservoirs, dams, streams/rivers, railway lines, and highways.

Environmental assessment refers to an understanding of the present status of the environment and the impacts on the environment because of the proposed activity, and a study of how to manage them. EIA is potentially one of the most valuable, interdisciplinary, objective decision-making tools with respect to alternate routes of development-process technologies and project sites. It is anticipatory mechanisms that establish quantitative values for parameters indicating the quality of environment before, during, and after the proposed development activity, thus allowing measures that ensure environmental compatibility. Collection of baseline information through intensive field monitoring is one of the crucial parts of EIA in view of the resources, time, and financial input required for it.

The EIA may be defined as the documentation of an environmental analysis, which includes identification, implementation, prediction, and mitigation of impacts caused by a proposed developmental activity.

Rapid EIA (REIA) refers to assessment based on one season (4 months) of monitoring for baseline data collection. Comprehensive EIA refers to assessment based on three seasons (12 months) of monitoring for the baseline data collection.

The purpose of the REIA report is to give a brief idea about the proposed project activities and their effects on the surrounding environment. Project proponents have to submit the REIA report to the Impact Assessment Agency (IAA), which is a committee formed from the State Pollution Control Board (SPCB) for the clearance of the projects, for their suggestions and recommendations about the project. Also, IAA will decide whether REIA is sufficient for the proposed project or whether a comprehensive environment impact assessment (CEIA) is required.

The CEIA means that the report contains premonsoon and postmonsoon season data, i.e., a comprehensive study on proposed project activities. Study of comprehensive EIA will take at least one year's time. The purpose of this is to assess the status of the environment around the proposed project location for attributes, e.g., air, water, noise, land use/land cover, socioeconomic and rehabilitation and resettlement factors, etc., to identify and quantify the impacts of various project operations by assigning environmental attributes, to evaluate the impacts on an environmental quality basis and prepare an environmental management plan (EMP) stipulating control measures to be adopted for mitigating the adverse impacts (if any) and to prepare a postproject environmental quality monitoring program.

9.2 ENVIRONMENT IMPACT ASSESSMENT STEPS

The various steps involved in the process of EIA are as shown in Fig. 9.1.

9.2.1 Screening

Screening is the first step of any EIA project. It provides information on whether a project is subjected to or excluded from an EIA. The screening criteria are based on scale of investment, type of development, and location of development.

9.2.2 Scoping

Scoping is the first task to determine the focus of EIA and to identify key impacts. The scope of a study should address all the issues of significance and importance that decision makers need to get the answer. Quantifiable impacts are to be assessed on the basis of magnitude, prevalence, frequency, and duration. EIA is ideally undertaken for a project and its alternatives (e.g., different locations, scale, design, etc.).

9.2.3 Baseline Studies

Baseline data is to be collected on land use, land cover, land environment, ambient air quality, water environment, biological environment, and

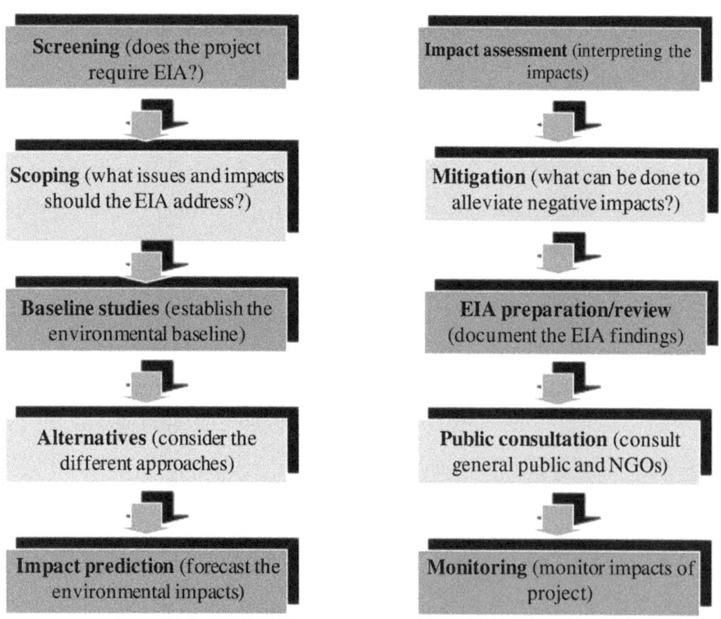

Figure 9.1 The steps of environment impact assessment (EIA).

socioeconomic factors. The baseline study involves gathering and evaluating information from existing sources and collecting field data. The existing sources of information (secondary data) may include databases, reports, and local community. Primary data is generated through field works, which include monitoring and surveys. The field monitoring and surveys includes meteorological parameters, ambient air quality, noise data, water quality (surface and ground water), land environment, and socioeconomic factors. The details are as given in Table 9.1.

Socioeconomic data is generated by designing a questionnaire that will provide the baseline data on people's status.

9.2.4 Impact Identification, Prediction, Assessment, and Evaluation

Environmental impact is identified in two categories, i.e., primary impacts and secondary impacts, and are assessed during the construction and operation phases. Various impacts attributed during the construction phase are directly related to employment during construction, such as possible influx of labor and stress on public utilities and services; use of water and power and its source during construction; proposed earth moving, dredging, and drilling operations; proposed plan for transportation and storage of construction material; detailed schedule of activity, resource requirements, and disposal of solid waste/dredged material.

Various impacts attributed during the operational phase are direct employment for operation; raw materials, fuels, and chemicals used, and their quantities, characteristics, and arrangement for transport to site, storage facilities, etc.; detailed manufacturing processes along with flow diagram; list of main equipment and machinery; built-in pollution control equipment and their efficiencies, etc.

The impact assessment for the proposed project with and without EMP is as given in Table 9.2.

9.2.5 Assessment Methodologies

Following completion of the evaluation of environmental parameters/effects (which will correspond to the feasibility stage of project planning), a project assessment can be formulated that:

1. lists each significant effect with its quantification;
2. compares these and their interrelationships;

Table 9.1 Attributes of field monitoring and surveys

Meteorology

Attributes	**Sampling**		**Measurement method**
Meteorological	**Network**	**Frequency**	
Wind speed, wind direction, dry bulb temperature, wet bulb temperature, relative humidity, rainfall, solar radiation, cloud cover, environmental lapse rate	Minimum one site in the project likely impact area	1 hourly continuous	Mechanical/ automatic weather stationRain gaugeAs per meteorology department specifications

Ambient air quality

Suspended particulate matter (SPM)	24 hourly twice a week	
Respirable suspended particulate matter (RSPM)	24 hourly twice a week	Measurement method: High volume sampler Monitoring network: Minimum two locations in upwind side, more sites in downwind side/ impact zone
Sulfur dioxide (SO_2)	8 hourly twice a week	
Oxides of nitrogen (NOx)	8 hourly twice a week	
Carbon monoxide (CO)	8 hourly twice a week	
Hydrogen sulfide (H_2S)	24 hourly twice a week	

Noise data

Attributes	**Sampling**		**Measurement method**
Noise	**Network**	**Frequency**	
Hourly equivalent noise levels	Identified study area	Once in each season	Instrument: noise level meter

(Continued)

Table 9.1 (Continued)

Noise data

Attributes	**Sampling**		**Measurement method**
Noise	**Network**	**Frequency**	
Hourly equivalent noise levels	In-plant (1.5 m from machinery)	Once	Instrument: noise level meter
Hourly equivalent noise levels	Highways	Once in each season	Instrument: noise level meter

Water quality

Attributes	**Sampling**		**Measurement method**
Parameters for water quality	**Network**	**Frequency**	
pH, temperature, turbidity, magnesium hardness, total alkalinity, chloride, sulfate, nitrate, fluoride, sodium, potassium, total nitrogen,total phosphorus, dissolved oxygen, biochemical oxygen demand, chemical oxygen demand, phenol, heavy metals, total coliforms, fecal coliforms, phytoplankton, and zooplankton	Set of grab samples during pre- and postmonsoon for ground and surface water for 10 km distance	Diurnal and seasonwise	Analyzed as per: • Prescribed by the regulatory authorities • Standard methods

Land environment

Attributes	Sampling		Measurement method
Parameters			
Soil	One surface sample from each village, (soil samples be collected as per BIS specifications)	Season wise	Collected and analyzed as per soil analysis (M.L. Jackson, 2005 and soil analysis reference book by C.A. Black, 1965)
Particle size distribution, texture, pH, electrical conductivity, cation exchange capacity, alkali metals, sodium absorption ratio (SAR), permeability, water holding capacity, and porosity			
Land use/land cover	At least 20 points along the boundary		Global positioning system; topo sheets; satellite imageries (1:25,000) and satellite imageries (1:25,000) (project specific)
Location code, total project area, topography, drainage (natural), cultivated, forest, plantations, water bodies, roads, and settlements			

Abbreviations: BIS, Bureau of Indian Standards.

3. derives conclusions on the overall environmental integrity of the project, and on feasible modification to the project plan that will serve to minimize adverse effects and/or to offset these with positive enhancement measures, or otherwise to optimize the overall project concept and resulting benefits within acceptable benefit/cost ratio limits;

Table 9.2 Impact assessment for the proposed project with and without EMP

Environmental components likely to be affected		EIA without EMP Project activities likely to affect environmental components								
		Site preparation	Construction/ other activities	Excavation	Drilling/Blasting	Dumping (Muck disposal)	Transportation	Waste water generation	Project completion	Total impact on component
Air quality	M									
	I									
Noise and vibration	M									
	I									
Surface-water quality	M									
	I									
Ground-water quality	M									
	I									
Soil quality/erosion	M									
	I									
Land-use pattern	M									
	I									
Flora and fauna	M									
	I									
Aesthetics	M									
	I									
Human health	M									
	I									
Socioeconomic status	M									
	I									
Economy, trade, and commerce	M									
	I									
Total action impact										

EIA With EMP

Environmental components Likely to be Affected		Site preparation	Construction/ other activities	Excavation	Drilling/Blasting	Solid waste disposal and reuse	Transportation	Waste water generation	Greenbelt development	Project completion	Total impact on component
		Project activities likely to affect environmental components									
Air quality	M										
	I										
Noise and vibration	M										
	I										
Surface-water quality	M										
	I										
Ground-water quality	M										
	I										
Soil quality/ erosion	M										
	I										
Land-use pattern	M										
	I										
Flora and fauna	M										
	I										
Aesthetics	M										
	I										
Human health	M										
	I										
Socioeconomic status	M										
	I										
Economy, trade, and commerce	M										
	I										
Total action impact											

I, Importance; M, Magnitude
Impact scale: 1, Minimal; 2, Appreciable; 3, Significant; 4, Severe.
Positive sign (or no sign), beneficial impact; negative sign, adverse impact; blank, no impact.

4. presents the recommended environmental monitoring program to be carded out as part of project operation and maintenance (O & M), together with delineation of tasks, skills needed, cost, procedures for report preparation and distribution, suggestions for participation by various interested agencies with cost sharing, and designation of the monitoring coordinator.

Numerous techniques and methods have been developed for evaluating and presenting the impacts of proposed and ongoing developmental activities on the environment. One of the common misconceptions is that only one of the general assessment methodologies is used in EIA. The fact is that for projects such as land clearance projects, which have very diverse and widespread impacts, the EIA team may wish to utilize several methods.

The EIA analyst, or the person charged with the preparation of an EIA report, is faced with a vast quantity of raw and usually unorganized data. Hence, each technique and method for the evaluation of impacts should have the following qualities and characteristics:

1. It should be systematic in approach.
2. It should be able to organize a large mass of heterogeneous data.
3. It should be able to quantify, relatively accurately, the impacts.
4. It should be capable of summarizing such data.
5. It should be able to aggregate the data into sets with the least loss of information because of the aggregation.
6. It should have a good predictive capability.
7. It should extract the salient features.
8. It should be able to, finally, display the raw data and the derived information in a meaningful fashion.

Each of the different methodologies for the assessment of environmental impacts of developmental projects do have their advantages and disadvantages and their utility for a particular application is largely a matter of choice and judgment of the analyst. Nevertheless, some objective criteria exist in making such a choice and these are stated below under the key areas that face the assessment process.

9.2.5.1 General

The methodology should be simple so that the available manpower with limited background knowledge can grasp and apply it without much difficulty. It should lend itself to being applied by a small group with a limited budget and under time constraints. It should be flexible enough to allow for modifications and changes through the course of the study, especially when more detailed examination is later required.

9.2.5.2 Impact Identification

The methodology should be sufficiently comprehensive to contain all possible options and alternatives and should give enough information on them to enable proper decision making. It should identify specific parameters on which there would be significant impacts. It should require and suggest methods for identifying project impacts as distinguished from future environmental changes produced by other causes. It should be able to identify accurately the location and extent of the impacts on a temporal scale.

9.2.5.3 Impact Measurement

The methodology should be a commensurate set of units so that comparison can be made between alternatives and criteria. It should suggest specific measurable indicators to be used to quantify impacts on the relevant environmental parameters. It should provide for the measurement of impact magnitude as distinct from impact significance. It should be based on criteria that are as objective as possible and that are stated explicitly.

9.2.5.4 Impact Interpretation and Evaluation

The methodology should be able to assess explicitly the significance of measured impacts on a local, regional, and national scale. The criteria and assumptions employed to determine impact significance should be explicitly stated. The effects on the environment with and without the project should be clearly portrayed so that decision makers can easily understand the exact nature of impacts caused by the project. It should be able to aggregate the vast amounts of information and raw input data. Uncertainty of possible impacts is a very real problem in EIA. The methodology should be able to take this aspect in account. It should identify impacts that have low probability of occurrence but a high potential for damage and loss. It should be such that the conclusions that are derived from it provide sufficient depth of analysis. It should provide a sufficiently detailed and complete comparison of the various alternatives readily available for the project under study. It should require and suggest a mechanism for public involvement in the interpretation of the impacts and their significance.

9.2.5.5 Impact Communication

The methodology should provide a mechanism for linking impacts to specific affected geographical or social groups. It should provide a description of the project setting to aid the users in developing an adequately comprehensive overall perspective. It should provide the results of the impact analysis summarized in a format that will give the users, who range from the lay public to the decision makers, sufficient detail to understand it and have confidence in its assessment. It should provide a format for highlighting the key issues and impacts identified in the analysis. One of the most important factors in choosing a methodology is whether it is able to comply with the tenses of reference established by the controlling agency.

Impact assessment methodologies are progressively changing from a static, piecemeal approach to one that reflects the dynamism of nature and the environment. Consequently, the trend is away from mere listing of potential impacts towards more complex modes whereby the methodologies can identify feedback paths, higher-order impacts than merely those apparent, first-order ones, and uncertainties. In short, the methodological trend is approaching an overall management perspective. There are a variety of techniques and methods to evaluate environmental impacts of development projects.

Among the more important techniques and methodologies useful for assessing the impacts of developmental activities on the environment in developing countries are ad hoc, checklist, matrices, networks, overlays, environmental index using factor analysis, cost/benefit analysis, and simulation modeling workshops. It is important to understand their drawbacks in order to determine which methods are most appropriate. An evaluation of various methodologies by Lohani et. al., is presented in Table 9.3.

9.2.5.6 Ad Hoc

The ad hoc method, while being a simple one that can be performed without any training, merely presents the pertinent information of a project's effects on the environment without any sort of relative weighting or any cause—effect relationships. It does not even state the actual impacts on specific parameters that will be affected.

The ad hoc method has the following drawbacks:

1. It gives no assurance that it encompasses a comprehensive set of all relevant impacts.
2. It lacks consistency in analysis as it may select different criteria to evaluate different groups of factors.
3. It is inherently inefficient, as it requires a sizeable effort in identifying and assembling an appropriate panel for each assessment.

Table 9.3 Summary of current EIA methodology evaluation

S.No.	Criteria	Checklists	Overlay	Network	Matrix	Environmental index	Cost/ benefit analysis	Simulation modeling workshop
1.	Comprehensiveness							
2.	Communicability							
3.	Flexibility							
4.	Objectivity							
5.	Aggregation							
6.	Replicability							
7.	Multifunction							
8.	Uncertainty							
9.	Space dimension							
10.	Time dimension							
11.	Data requirement							
12.	Summary format							
13.	Alternative comparison							
14.	Time requirement							
15.	Manpower requirement							
16.	Economy							

L, Completely fulfilled, or low-resource need; S, Partially fulfilled, or moderate resource need; N, Negligibly fulfilled, or high-resource need.
Source: Environmental Impact Assessment: Guidelines for planners and decision makers, UN Publication ST/ESCAP/351, ESCAP, 1985.

Because of the above drawbacks, it is not recommended as a method for impact analysis. It is as the name indicates, an ad hoc method, and it has utility only when other methods cannot be used because of lack of expertise, resources, etc.

9.2.5.7 Checklists

Checklists, in general, are strong in impact identification and are capable of bringing impacts to the attention and awareness of their audiences. Impact identification is the most fundamental function of an EIA and in this respect all types of checklists, including simple, descriptive, scaling, and weighting checklists, do well. But simple and descriptive checklists offer no more than just this. They merely identify the possible potential impacts without any sort of rating as to their relative magnitudes. As a result they are most applicable at the Initial Environmental Evaluation (IEE) stage of an assessment.

The Oregon method goes a step further than this and provides an idea of the nature of the impact by means of assigning a textual rating of the impact as long-term, direct, etc. But nevertheless, this approach is not suitable for impact measurement and does not aid much in the decision-making process. Rather, it identifies the impacts and leaves the interpretation to the decision makers.

The element of scaling and weighting that is inherent in the latter types of checklists lends itself move easily to decision making. Such checklists, apart from being strong in impact identification, also incorporate the functions of impact measurement and to a certain degree, those of interpretation and evaluation, and it is those aspects that make them more amenable for decision-making analysis.

9.2.5.7.1 Sample Modified Oregon Checklist for Reservoir Project

(Instructions):

Answer the following questions by placing an "X" in the appropriate YES/NO space, considering activity, construction, operational, as well as indirect impacts.

Use the "explanation" section to clarify points or add information.

9.2.5.7.2 Natural Biological Environment

1. Might the proposed activity affect any natural feature or water resource adjacent to or near the activity areas? NO X YES

If YES, specify natural feature affected.

	Direct	Indirect	Synergistic	Short-term	Long-term	Reversible	Irreversible	Severe	Moderate	Insignificant
1 Surface-water hydrology	(X)	( )	( )	( )	(X)	( )	(X)	( )	( )	(X)
2 Surface-water quality	(X)	( )	( )	( )	(X)	( )	(X)	(X)	( )	( )
3 Soil/erosion	(X)	( )	( )	( )	(X)	( )	(X)	(X)	( )	( )
4 Geology	(X)	( )	( )	( )	(X)	( )	(X)	(X)	( )	( )
5 Climate	(X)	( )	( )	( )	(X)	( )	(X)	(X)	( )	( )

2. Might the activity affect wildlife or fisheries ? NO X YES
If YES, specify wildlife or fisheries affected:

1	Wildlife habitat	(X)	( )	( )	( )	(X)	( )	(X)	(X)	( )	( )
2	Ecology of fisheries	(X)	( )	( )	( )	(X)	( )	(X)	(X)	(X)	( )

3. Might the activity affect natural vegetation? NO X YES
If YES, specify vegetation and area affected.
4. Are there any other developments planned that are or will be impacted by the proposed activity, including those beyond the control of the submitting agency? NO X YES
If YES, specify other development affected.

Source: Application and evaluation of EIA methodologies using a small dam/reservoir case study. Presented to the Intercountry Workshop on Rapid Techniques for Environmental Assessment in Developing Countries. Asian Institute of Technology, Bangkok, February 1–5, 1982.

But impact scaling and weighting is, nevertheless, subjective, and this poses the danger that society holds all diverse impacts to be equally important. Further, it implicitly assumes that numerical values assigned to impacts can be derived on the basis of expert knowledge and judgment alone. Both these suppositions are suspect on deeper introspection.

The Adams–Burke and the Odum optimum-pathway methods provide subjective estimates of the impacts and this subjectivity, although somewhat compensated for in the latter by the introduction of an error term in the calculation of the index to account for misjudgment of relative impacts, reduces the utility of these methods and the replicability of the analysis. The approaches become valid only for a case-by-case comparison of alternatives.

Methods that involve scaling and weighting and the consequent aggregation remove decision making from the hands of decision makers. Further, they incorporate into one number various intrinsically different impacts and this deprives the decision maker of the possibility of tradeoffs.

The water resources assessment methodology (WRAM) was developed to overcome such limitations in the aggregation methods. It offers greater flexibility and adaptability to local conditions and preferences and gives the option of making tradeoffs in the decision-making process to the appropriate decision makers.

But like other checklist-based methods, WRAM does not convey the dynamic nature of impacts, nor does it provide meaningful information regarding higher-order effects and impacts and interactions between impacts to its audiences.

9.2.5.8 Matrices

Matrices provide cause–effect relationships between the various project activities and their impacts on the numerous environmentally important sectors or components. Matrices provide a graphic tool for displaying impacts to their audience in a manner that can be easily comprehended. Simple matrices, while able to identify first-order effects, cannot show higher interactive effects between impacts. Simple interaction matrices largely overcome this limitation. But such matrices are generally useful for depicting ecological interactions only and for the sake of documentation. While the scale of the interaction is identified, individual actions of the project are not correlated with their resulting impacts on the environmental components.

Graded and quantitative matrices offer that element of rating and weighting that is lacking in the simple interaction matrices, and through this offer criteria for decision-making analysis. However, the most serious criticism of such weighting matrices, and this can also be extended to scaling and weighting checklists, is that through the inherent aggregation process, decision making is, in effect, removed from the hands of the decision makers and the concerned public. A great deal of information that is valuable to decision making is lost in the conversion to numbers.

This loss of information has been elucidated as follows:

1. Weights are assigned to environmental components and consequently to impacts, without any guarantee that such weights and ratings will represent the actual impacts that will be apparent once the project is implemented and operational.
2. What is generally called an objective procedure, the assignment of weights and the subsequent quantification, is in fact an arbitrary assignment of scales of "environmental quality" based on the value judgment of "experts."
3. Aggregation of numerical impacts through suitable transformation functions results in the combination of inherently different items into a single index or number and leads to loss of information about the various impacts from the numerous project actions, thereby precluding the possibility of tradeoffs by decision makers.

Matrices are strong in identifying impacts and, unlike checklists, can also represent higher-order effects and interactions. Since the dynamic nature of impacts can also be identified, they can also provide the functions of impact measurement, interpretation and evaluation and can communicate the results in an easily understood format to their audiences. However, they cannot compare alternatives in a single format and different alternatives need to be assessed and presented separately.

9.2.5.9 Networks

Networks are capable of identifying direct and indirect impacts, higher-order effects and interactions between impacts and, hence, are able to identify and incorporate mitigation and management measures into the planning stages of a project. They are suitable for expressing ecological impacts but of lesser utility in considerations involving social, human, and esthetic aspects. This is because weightings and ratings of impacts are not features of network analysis. Networks generally consider only adverse impacts on the environment and hence decision making in terms of the cost and benefit of a development project to a region is not amenable to network analysis. Temporal considerations are not properly accounted for and short-term and long-term impacts are not differentiated to the extent required for easy understanding.

While networks can incorporate several alternatives into their format, the display becomes very large and hence unwieldy when large regional plans are being considered. Further, networks are capable of presenting scientific and factual information, but provide no avenue for public participation.

9.2.5.10 Overlays

Overlays are useful when addressing questions of site and route selection. They provide a suitable and effective mode of presentation and display to their audiences. But overlay analysis cannot be the sole criterion for EIA.

There is no provision for quantification and measurement of the impacts or for coverage of all impacts. The considerations in overlay analysis are purely spatial, and temporal considerations are outside its scope. Social, human, and economic aspects are not accorded any consideration. Further, higher-order impacts cannot be identified.

Overlays are very useful for Land cover (LC) project EIAs for comparing land capabilities, existing and projected land use, road route alternatives, and other similar parameters.

9.2.5.11 Environmental Index Using Factor Analysis

A factor analysis-based environmental index approach for assessing impacts is a useful technique in that it enables the grouping and clustering of parameters and components that are complex in nature. However, factor analysis is useful only when there is some sort of meaningful variation in the dataset on hand. Without this, only one factor can be derived from a dataset and the factor analysis exercise loses its significance and potential.

Factor analysis identifies components that will be adversely affected by certain project activities and can provide a fair degree of measurement capability of the impact. But factor analysis requires proper interpretation and evaluation of the results by the analyst to be of utility. The results from factor analysis are not, on their own, amenable to interpretation.

The index based on factor analysis provides criteria for decision-making analysis. It can consider all types of impacts on all types of components. A feature of factor analysis is that it gives the relative amount of variability that is explained by the factors. This is useful in ascertaining the relative degree of magnitude and scale of an impact on a component. But the criticism directed towards aggregating methods that was mentioned earlier is also true in this case.

9.2.5.12 Cost/Benefit Analysis

Cost/benefit analysis provides the nature of expense and benefit accruable from a project in monetary terms as is common practice in traditional feasibility studies, and hence enables easy understanding and aids decision making tremendously.

The difficulty encountered in the use of this technique has been that impacts have to be transformed and stated in explicit monetary terms. This is not always possible, especially for intangibles like the monetary value of the damages to health owing to the advent of cholera, etc.

Cost/benefit analysis of the type for assessment of natural systems is not merely concerned with the effects on environmental quality, but rather, it seeks the conditions for sustainable use of the natural resources in a region. This type of approach is not useful for small-scale development projects, but is better suited for the analysis and evaluation of a regional development plan.

9.2.5.13 Simulation Modeling Workshops

System analysts have developed an approach to EIA and management commonly referred to as adaptive environmental assessment and management (AEAM), which combines various simulation models to predict impacts. This approach broadens the potential of simulation models to evaluate the impacts of alternatives and is beneficial for project planning. It was used by the Secretariat of the Interim Committee for Coordination of Investigations of the Lower Mekong Basin in developing and testing their EIA guidelines for tropical river-basin development.

The AEAM approach uses small interdisciplinary teams interacting through modeling workshops over a relatively short time to predict impacts and evaluate alternatives including management measures. The adaptive assessment process can be divided into three types of workshops:

- the initial workshop
- the second-phase workshop
- the transfer workshop

The AEAM technique largely overcomes the shortcomings of most other methods in that other methods assume unchanging conditions or project impacts in a single time frame on statistically described environmental conditions. It also overcomes a built-in bias towards compartmentalization and fragmentation of the relationship between project actions, environmental characteristics, and likely impacts, while the reality may be that the actual impacts may alter the scale and direction of change within the environmental and social system. The AEAM technique enables the assessment of development projects in the light of the reality of such changes.

The AEAM technique can handle higher-order impacts and interactions between impacts. But it depends on a small group of experts and has no avenue for public participation. This aspect is of particular significance for large-scale development, where the opinions of interest groups are important.

9.2.6 Environmental Management Plan

An EMP describes the management systems, monitoring mechanisms, and the auditing arrangements required to ensure both the proper implementation of agreed mitigation and the verification of predicted environmental impacts. An EMP aims at controlling pollution at the source level to the possible extent with the available and affordable technology followed by treatment before it is discharged, and the preservation of ecological systems by considering built-in pollution abatement facilities at the proposed site.

9.2.6.1 Greenbelt Development

Greenbelt development has been recommended as one of the major components of EMP, and will improve ecology, environment, and quality of the surrounding areas through mitigation of fugitive emissions, attenuation of noise levels, waste-water reuse, development of ecosystems, creation of an esthetic environment, and use of waste land to improve environmental quality.

9.2.6.2 Financial Plan for Implementing EMP

A financial plan should be drawn that should include the annual expenditures for the next 5 years for implementation of the EMP. The expenditure estimated in the financial plan of the EMP has to be included in the detailed feasibility/project report of the project.

9.2.7 Public Consultation

The public must be informed and consulted on a proposed development after the completion of the EIA report. The affected persons may include local residents, local associations, any other persons located at the project site/sites of displacement, etc. They are to be given an opportunity to make oral/written suggestions to the SPCB.

Public consultation refers to the process by which the concerns of local affected persons and others who have a plausible stake in the environmental impacts of the project or activity are ascertained with a view to taking into account all the material concerns in the project or activity design as appropriate. All Category A and Category B1 projects or activities shall undertake public consultation, except (1) modernization of irrigation projects, (2) all projects or activities located within industrial estates or parks, (3) expansion of roads and highways, (4) all building/construction projects/area development projects and townships, (5) all category B2 projects and activities, and (6) all projects or activities concerning national defense and security or involving other strategic considerations as determined by the central government.

The public consultation shall ordinarily have two components comprising of (1) a public hearing at the site or in its close proximity districtwise for ascertaining concerns of local affected persons, and (2) obtaining responses in writing from other concerned persons having a plausible stake in the environmental aspects of the project or activity. This is to be conducted by the SPCB or the Union Territory Pollution Control Committee (UTPCC) concerned in the specified manner and the proceedings should be forwarded to the regulatory authority concerned within 45 days (45 days from a request to the effect from the applicant).

After completion of the public consultation, the applicant shall address all the material environmental concerns expressed during this process, and make appropriate changes in the draft EIA and EMP. The final EIA report, so prepared, shall be submitted by the applicant to the concerned regulatory authority for appraisal. The applicant may alternatively submit a supplementary report to draft EIA and EMP addressing all the concerns expressed during the public consultation.

9.2.7.1 Appraisal

Appraisal means the detailed scrutiny by the Expert Appraisal Committee or State Level Expert Appraisal Committee of the application and other documents like the final EIA report, outcome of the public consultations including public hearing proceedings, submitted by the applicant to the regulatory authority concerned for grant of environmental clearance. This appraisal shall be made by Expert Appraisal Committee or State Level Expert Appraisal Committee concerned in a transparent manner in a proceeding to which the applicant shall be invited for furnishing necessary clarifications in person or through an authorized representative. On conclusion of this proceeding, the Expert Appraisal Committee or State Level Expert Appraisal Committee concerned shall make categorical recommendations to the regulatory authority concerned either for grant of prior environmental clearance on stipulated terms and conditions, or rejection of the application for prior environmental clearance, together with reasons for the same.

Validity of environmental clearance is a maximum of 30 years for mining projects, 10 years for river valley projects, 5 years for all other projects, limited period for area development projects till the developer is responsible and can be extended to another 5 years upon submission of application within validity period (Table 9.4).

9.3 LEGAL PROVISION

EIA notification 1994 and subsequent amendments, under EP (Act), 1986, (Published in the Gazette of India, extraordinary, Part-II, and Section 3, Subsection (ii) by Ministry of Environment and Forests, New Delhi, September 14, 2006. However, the legal provisions change from time to time depending on the minor amendments made in the Rules for handling and management of wastes.

Table 9.4 List of projects or activities requiring prior environmental clearance

Project or activity		Category with threshold limit		Conditions if any
		A	B	
1.		Mining, extraction of natural resources, and power generation (for a specified production capacity)		
(1)	(2)	(3)	(4)	(5)
1(a)	Mining of minerals	$\geq$50 ha of mining lease area. Asbestos mining irrespective of mining area	<50 ha $\geq$5 ha of mining lease area	General condition shall apply. Note: mineral prospecting (not involving drilling) is exempted provided the concession areas have gotten previous clearance for physical survey
1(b)	Offshore and onshore oil and gas exploration, development and production	All projects		Note: exploration surveys (not involving drilling) are exempted provided the concession areas have gotten previous clearance for physical survey
1(c)	River valley projects	1. $\geq$50 MW hydroelectric power generation; 2. $\geq$10,000 ha of culturable command area	1. <50 MW $\geq$25 MW hydroelectric power generation; 2. <10,000 ha of culturable command area	General condition shall apply

1(d)	Thermal power plants	1. ≥ 500 MW (coal/lignite/naphtha and gas based); 2. ≥ 50 MW (pet coke, diesel, and all other fuels)	1. < 500 MW (coal/lignite/naphtha and gas based); 2. < 50 MW ≥ 5 MW (pet coke, diesel, and all other fuels)	General condition shall apply
1(e)	Nuclear power projects and processing of nuclear fuel	All projects	–	
2.		Primary processing		
2(a)	Coal washeries	≥ 1 million tonne/annum throughput of coal	<1million tonnne/annum throughput of coal	General condition shall apply (if located within mining area the proposal shall be appraised together with the mining proposal)
2 (b)	Mineral beneficiation	≥ 0.1 million tonne/annum mineral throughput	< 0.1million tonne/annum mineral throughput	General condition shall apply (mining proposal with mineral beneficiation shall be appraised together for grant of clearance)
3.		Materials production		
3(a)	Metallurgical industries (ferrous and nonferrous)	1. Primary metallurgical industry (all projects)	Sponge iron manufacturing <200 TPD secondary	General condition shall apply for sponge iron manufacturing

(*Continued*)

Table 9.4 (Continued)

Project or activity		Category with threshold limit: A	B	Conditions if any
		2. Sponge iron manufacturing ≥200 TPD 3. Secondary metallurgical processing industry. All toxic and heavy metal producing units ≥20,000 tonnes/annum	metallurgical processing industry 1. All toxic and heavy metal producing units <20,000 tonnes/annum 2. All other nontoxic secondary metallurgical processing industries >5000 tonnes/annum	
3(b)	Cement plants	≥ 1.0 million tonnes/annum production capacity	<1.0 million tonnes/annum production capacity. All standalone grinding units	General condition shall apply
4.		Materials processing		
4(a)	Petroleum refining industry	All projects	–	–
4(b)	Coke oven plants	≥250,000 tonnes/annum	<250,000 and ≥ 25,000 tonnes/annum	–
4(c)	Asbestos milling and asbestos-based products	All projects	–	–
4(d)	Chlor-alkali industry	≥300 TPD production capacity or a unit located outside the notified industrial area/estate	<300 TPD production capacity and located within a notified industrial area/estate	Specific condition shall apply. No new mercury cell-based plants will be permitted and existing

				units converting to membrane cell technology are exempted from this notification
4(e)	Soda ash industry	All projects	–	–
4(f)	Leather- /skin- /hide-processing industry	New projects outside the industrial area or expansion of existing units outside the industrial area	All new or expansion of projects located within a notified industrial area/ estate	Specific condition shall apply
5.		Manufacturing/fabrication		
5(a)	Chemical fertilizers	All projects	–	–
5(b)	Pesticides industry and pesticide specific intermediates (excluding formulations)	All units producing technical grade pesticides	–	–
5(c)	Pulp and paper industry excluding manufacturing of paper from waste paper and manufacture of paper from ready pulp without bleaching	Pulp manufacturing and pulp and paper manufacturing industry	Paper manufacturing industry without pulp manufacturing	General condition shall apply
5(d)	Sugar industry	–	≥5000 tcd cane crushing capacity	General condition shall apply
5(e)	Induction/arc furnaces/ cupola furnaces 5TPH or more	–	All projects	General condition shall apply

(Continued)

Table 9.4 (Continued)

Project or activity		Category with threshold limit		Conditions if any
		A	B	
6.		Service sectors		
6(a)	Oil and gas transportation pipe line (crude and refinery/petrochemical products), passing through national parks/sanctuaries/coral reefs/ecologically sensitive areas including LNG Terminal	All projects		–
(1)	(2)	(3)	(4)	(5)
6(b)	Isolated storage and handling of hazardous chemicals (as per threshold planning quantity indicated in column 3 of schedule 2 & 3 of MSIHC Rules 1989 amended 2000)	–	All projects	General condition shall apply
7.		Physical infrastructure including environmental services		
7(a)	Airports	All projects	–	–
7(b)	All ship-breaking yards including ship-breaking units	All projects	–	–

7(c)	Industrial estates/parks/ complexes/ areas, export-processing zones (EPZs), special economic zones (SEZs), biotech parks, leather complexes	If at least one industry in the proposed industrial estate falls under Category A, the entire industrial area shall be treated as Category A, irrespective of the area. Industrial estates with area greater than 500 ha and housing at least one Category B industry	Industrial estates housing at least one Category B industry and area <500 ha Industrial estates of area $>$ 500 ha and not housing any industry belonging to Category A or B	Special condition shall apply. Note: industrial estate of area below 500 ha and not housing any industry of Category A or B does not require clearance
7(d)	Common hazardous waste treatment, storage and disposal facilities (TSDFs)	All integrated facilities having incineration and landfill or incineration alone	All facilities having land fill only	General condition shall apply
8.		Building/construction projects/area development projects and townships		
8(a)	Building and construction projects		$\geq$ 20,000 sq.m. and $<$150,000 sq.m. of built-up area	Built-up area for covered construction; in the case of facilities open to the sky, it will be the activity area
8(b)	Townships and area development projects		Covering an area $\geq$ 50 ha and or built up area $\geq$ 150,000 sq. m.	All projects under Item 8(b) shall be appraised as Category B1

CHAPTER 10

Landfill Gases

10.1 GREENHOUSE GASES AND CLIMATE CHANGE

Any gas that is capable of absorbing and emitting infrared radiation and is responsible for the greenhouse effect is known as a greenhouse gas (GHG). Over the years, rapid urbanization and industrial activities have increased the amount of GHGs like carbon dioxide, methane, nitrous oxide, and several others in the atmosphere. Increased radiative forcing from GHGs has changed the earth's reflectivity, which is responsible for the rise in temperatures around the globe. Fig. 10.1 shows the global percentage of different GHGs in 2010. While CO_2 contributes to 76%, CH_4, and N_2O respectively contributed to 16% and 6% global GHGs around the world. Other gases are mainly hydrofluorocarbons, perfluorocarbons, and sulfur hexafluoride.

Even though methane has lower concentrations and shorter lifetime than CO_2, it can entrap radiation more effectively than CO_2. Global methane emissions in 2005 from different sectors are shown in Fig. 10.2. Methane emissions from landfills are next only to the energy and agriculture sector.

10.2 LANDFILL GAS GENERATION

In developing countries like India, it is a common practice to dispose of generated solid waste in open areas without any prior treatment. Gases are emitted from landfills due to bacterial decomposition, volatilization, and chemical reactions. Of these, bacterial decomposition leads to maximum production of landfill gases.

Based on the nature of the waste and microbial action on organic matter present, landfill gases are produced. The major portion of landfill gas production is mainly due to the microbial action of methanogenic bacteria (methanogenesis). In a typical landfill, the formation of principal landfill gases like methane and carbon dioxide starts few months after the disposal due to degradation of waste, and reaches a maximum in $5-6$ years.

Solid and Hazardous Waste Management.
DOI: http://dx.doi.org/10.1016/B978-0-12-809734-2.00010-9

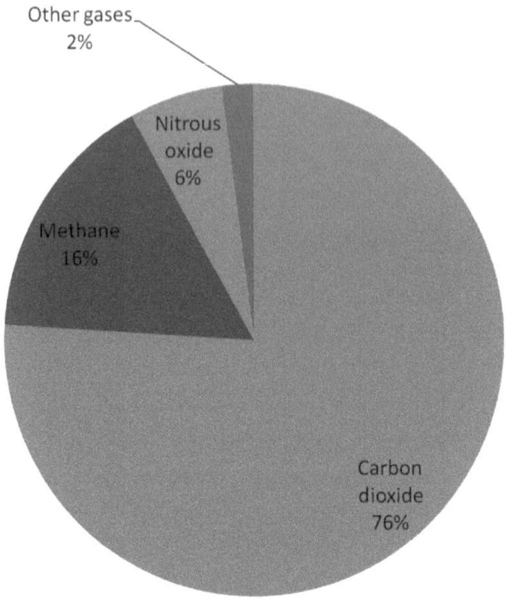

Figure 10.1 Greenhouse gas emissions by gas in 2010.

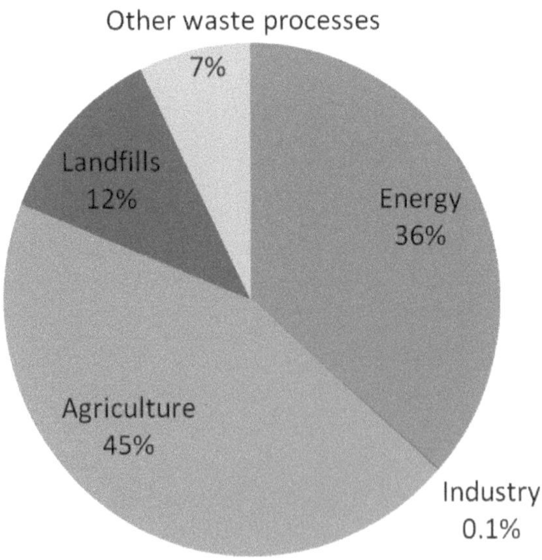

Figure 10.2 Global methane emissions from different sectors in 2005.

10.2.1 Phases of Bacterial Decomposition of Landfill

The generation of principal landfill gases occurs in five different sequential phases of bacterial decomposition as explained in Table 10.1.

Table 10.1 Phase wise generation of principal landfill gases

Phase	Description	Principal gases
I (Initial aerobic phase)	• The duration of this phase depends on the amount of trapped air in the landfill • Aerobic bacteria break down proteins, lipids, and long molecular chains of complex carbohydrates • Primary byproduct: CO_2	N_2, O_2, and CO_2
II (Acidic phase)	• Anaerobic bacteria consume compounds created in phase I to form alcohols, acetic, lactic, and formic acids • Primary byproduct: CO_2 and H_2	CO_2, N_2, and H_2
III (Neutral phase)	• Anaerobic bacteria consume organic acids produced in phase II to form acetate • The landfill is converted into neutral environment, and methanogenic bacteria try to establish themselves by having a symbiotic relationship with acid-generating bacteria • Primary byproduct: CO_2 and CH_4	CO_2, CH_4, N_2, and H_2
IV(Methane fermentation phase)	• These methanogenic bacteria continue to convert acetic acid and hydrogen gas into methane and carbon dioxide • Acid production rate decreases and methanogenesis becomes more dominant • Primary byproducts: CH_4 (45 − 60%) and CO_2 (40 − 60%).	CH_4, CO_2, and N_2
V (Maturation phase)	• Conditions shift from active degradation to dormancy • Landfill gas production drops • Depending on efficiency of capping, oxygen can reappear • Primary byproducts: CH_4, CO_2, and humic acid	CH_4, CO_2, N_2, and O_2

10.2.2 Factors Affecting Landfill Gases Generation

Landfill gas composition and its rate of production depend on waste characteristics and several environmental factors like presence of oxygen in landfill, temperature, and moisture content.

10.2.2.1 Composition of Waste

The composition and age of disposed waste plays a vital role in landfill gas production. The greater the organic content in the waste, the higher is the landfill gas production due to biodegradation. Additionally, the presence of nutrients such as sodium, potassium, etc., which help bacteria, also increase the production of landfill gases. In contrast, the presence of high salt concentrations in waste can inhibit the methane production.

10.2.2.2 Presence of Oxygen in the Landfill

The more the oxygen in the landfill, the longer the influence of aerobic bacteria and phase I of bacterial decomposition will be. If the waste in not compacted properly, the chances of oxygen in the voids are higher, which also leads to lower methanogenesis.

10.2.2.3 Moisture Content

Moisture content present in the landfill encourages bacterial activity and enhances the transport of nutrients and bacteria throughout the landfill. In densely packed landfills, the rate at which water infiltrates decreases. More water content is also useful in chemical reactions, which results in landfill gases. A moisture content of 40% leads to maximum gas production in a typical landfill.

10.2.2.4 Temperature

As the temperature increases, the landfill gas production increases due to an increase in bacterial activity. Temperature also increases rates of volatilization and chemical activities. Temperature is more significant in shallow landfills than deep landfills.

10.2.2.5 Age of Refuse

Newly buried waste produces more gases than older waste. Peak emissions occur within 5–7 years of waste burial. After 20 years, almost all the gas will have been produced, but emissions continue until 50 years postburial.

10.3 ON-FIELD AND LABORATORY TECHNIQUES TO ESTIMATE METHANE FROM LANDFILL

10.3.1 Stoichiometric Approach

This method is the simplest theoretical method used to estimate methane emissions. In this method, the chemical formula of waste in a landfill is postulated by using the analysis based on waste composition. By assuming total conversion of components, the yield of gas calculated by using following stoichiometric equation:

$$C_nH_aO_bN_c + [n - (a/4) - (b/2) + 3(c/4)]H_2O \rightarrow [(n/2) - (a/8) + (b/4) + 3(c/8)]CO_2 + cNH_3 + [(n/2) + (a/8) - (b/4) - (c/8)]CH_4$$

10.3.2 Flux Chamber Method

The flux chamber technique is a widely used, simple, and economical method to measure the emissions from the emitting source. This method uses a closed chamber with an open end used to trap the gas as it leaves the landfill surface. Right before the sampling, the top layer of the surface is to be excavated and the chamber is anchored at that point.

There are two types of flux chamber method: static closed chamber and dynamic closed chamber. In the static chamber method, the chamber is fixed on the study location and the gas emissions are allowed to build up within the chamber for a reasonable amount of time. The gas sample is drawn out of the chamber at equal intervals and saved in a glass vial for analysis in the laboratory. Enough care is to be taken that except for gas emitted from the soil surface, no removal or entry of gas into the chamber is allowed. Methane flux is calculated by measuring the change in concentration over the sampling time.

However, in the dynamic closed chamber method, air is allowed to flow through the chamber throughout the sampling time. Flux is calculated by measuring the methane concentration in the exhaust gas and air-flow rate. Overall, in both methods, the number of chamber tests conducted in the sampling area over the sampling period determines the accuracy of estimation.

10.3.3 Tracer-Gas Correlation Method

This method is a high-performance yet simple technique to measure the emission rates from wide-area sources like agricultural fields, sewer plants, and landfills. This method uses a tracer gas and fast-response spectroscopic instrument to measure the methane from the landfills. Pure acetylene or nitrous oxide is commonly used as tracer gas in this technique.

The tracer-gas correlation method (TGCM) follows two approaches of tracer gas correlation measurements, i.e., mobile and stationary. In both cases, using mass flow—controlling systems, tracer gas is released from the landfill surfaces, and using optical remote sensing, concentrations of both methane and tracer gas are measured. In the mobile approach, a cavity ring-down laser is fixed on a vehicle along with the arrangement for gas sample collection. In addition, a global positioning system and compact weather station are integrated to record the study of precise locations and meteorological data. Tracer gases with known emission rates are released from different parts of the landfill to measure the emission rates of landfill gases.

In stationary approach, long-path Fourier transform infrared spectroscopy (FTIR) is set up at a downwind location of the plume and tracer gas is emitted at a known emission rate. By using the known tracer-gas flux, methane emission rates from the landfill area are calculated. The total methane emission rates from the landfill are calculated as the product of tracer-gas emission rate and ratio of measured tracer-gas concentration and methane concentration above the background in the plume.

Similarly, in the mobile approach, using simultaneous measurements at different locations with controlled tracer-gas emission and meteorological data, the methane emission rates from the spatially inhomogeneous landfills can be estimated.

10.3.4 Micrometeorological Method

This method is relatively economical for measuring the methane emissions from landfills. It is a widely used method to measure emissions from the large flat-terrain areas with an assumption of uniform emissions over the surface. The micrometeorological method for measurement of methane emissions is based on the assessment of turbulent eddy currents in the vertical direction within the internal boundary layer.

Micrometeorological methods include techniques like flux gradient, eddy variance technique, integrated horizontal flux, and boundary layer budgeting. All these techniques are viable for the measurement of continuous emission data without any interceptions. The flux gradient and eddy variance techniques are the most commonly used micrometeorological techniques to measure the turbulent fluxes within the boundary layers. These techniques involve instrumentation facilities, mostly ultrasonic Doppler anemometers, to analyze the properties of the air parcel (pressure, temperature, humidity, wind speed) and concentrations of preferred gases.

10.3.5 Differential Absorption LiDAR

The differential absorption technique is an active and important technique to detect and measure the atmospheric concentrations of methane from landfills. In addition to methane, the differential absorption LiDAR (DIAL) method is often used to measure the ambient concentrations of gases such as ozone, H_2O, NO_2, CO, and SO_2, as well as particulate matter. The DIAL technique is widely used if the sample volume is less and it cannot be limited by the closed enclosure. In addition to landfill plumes, LiDAR techniques are used to measure the concentration of species from point sources like industrial facilities, and area sources like agricultural facilities.

This technique uses lasers to measure the contaminant concentrations in the area of study. Based on target constituents to be measured, a laser with suitable characteristics is selected. The lasers used may have wavelengths ranging from UV to IR. In this method, a laser beam is sent out into the atmosphere and a small portion of pulse is back-scattered to the detector. The time taken by the pulse to reach the detector is used to estimate the location of the measured concentration of methane. A two-dimensional concentration map can be generated using data from multiple lines of sight of the laser beam. A vertical scan of landfill plume in addition to wind speed is used to estimate methane flux.

10.3.6 Vertical Radial Plume–Mapping Method

This method is developed by the U.S. Environmental Protection Agency (EPA) to estimate the gaseous emissions from the fugitive-area sources like landfills. This method uses optical remote-sensing techniques, i.e., either open-path tunable diode laser or open-path FTIR to estimate the methane flux rates from the landfill surface. An optical remote-sensing system emits radiation at the study location and the reflected light is received by the detector, and the methane concentration is measured. In addition to the concentrations along the path, meteorological data is also collected. Path-integrated concentrations at different heights in different paths are calculated. This method consists of an algorithm that utilizes the measured concentrations, wind speed, and wind direction to calculate the methane flux across the plane.

10.4 MODELS USED TO ESTIMATE METHANE EMISSIONS FROM LANDFILLS

Empirical models used in different countries to estimate methane emissions from landfills can be broadly divided into two types based on rate of methane generation.

10.4.1 Zero-Order Models

In these models, the rate of methane emissions from landfills is assumed to be constant.

10.4.1.1 European Pollutant Emission Registrar: Germany

This model takes only unconditioned household waste into account. The amount of methane generated (α_t) in a year can be estimated using the equation below:

$$\alpha_t = 0.665\varsigma RAB$$

Where

ς = dissimilation factor; i.e., fraction of biodegradable carbon converted

A = amount of waste placed in a landfill (Mg waste/year)

B = fraction of biodegradable carbon in waste (Mg C/Mg waste)

R = factor depending on methane collected/recovered

0.1 (with active landfill-gas recovery and cover)

0.4 (active degreasing)

0.9 (no recovery)

10.4.1.2 Intergovernmental Panel on Climate Change Model

This model estimates amount of waste generated and its composition does not change over the years. The model was developed based on mass balance of waste generated from all the households in a locality.

$$\alpha_t = 0.66 W_t W_f FB\varsigma R(1 - O_x)$$

W_t = total municipal solid waste generated (Mg/year)

W_f = fraction of generated municipal solid wastes reaching landfills

F = methane correction factor

O_x = oxidation factor

10.4.2 First-Order Models

In these models, the decay of organic matter in the solid waste is assumed to follow the first order. Most of the commonly used models around the world are first order. Some of the prominent first-order decay models are as follows.

10.4.2.1 Triangular Method

In the absence of historic data about waste quantities, composition, and disposal practices, the triangular method can be used to estimate methane emissions from landfills. The triangular method results in methane emissions based on waste deposition variation and type of decomposition, i.e., aerobic or anaerobic. In this, total gas yield is computed for the organic fraction of waste and the biogas release is based on the first-order decay.

This method assumes that the rate of methane production is linear with time till it reaches a peak and falls linearly thereafter. In other words, in this method it is assumed that waste degradation process takes place in two phases. In the first phase, degradation starts after 1 year of deposition and increases until the peak is reached, and the second phase starts when the landfill gas generation rate starts decreasing and becomes zero. Thus, the area of the triangle is equivalent to the gas emitted over the time period by the amount of waste deposited.

10.4.2.2 Netherlands Single-Phase Model

This model assumes that the amount of waste disposed in landfills decomposes exponentially with time. Organic matter present in waste is assumed to decay exponentially over time. Quantification of methane emissions in this model done by using the following equation:

$$\alpha_t = \varsigma 1.87 A C_0 k\, e^{-kt}$$

Where

α_t = landfill gas formation at a specific time (m^3/year)

A = amount of waste placed (Mg)

C_o = organic carbon amount in the waste placed (kg C/Mg waste)

k = degradation rate constant (per year)

t = time elapsed since waste depositing (year).

10.4.2.3 Netherlands Multiphase Model

In a multiphase model, the composition of waste is considered, as different categories of waste degrade at different rates. This model considers eight waste categories, namely contaminated soil, construction and demolition, shredder waste, street-cleansing waste, sewage sludge and compost, coarse household waste, commercial waste, and household waste. The overall organic content in waste is classified based on the fraction of rapidly, moderately, and slowly degradable fraction of the waste. Methane production calculated by using following equation:

$\alpha_t = \varsigma \Sigma_{i=1}^{3} 1.87 A_i C_i K_i e^{-k_i t}$Where

α_t = landfill gas production at a specific time (m^3/year)

i = waste fraction type with degradation rate $k_{1,i}$ (kg_i/kg_{waste})

A_i = amount of waste placed (Mg)

C_i = organic carbon amount in the waste placed (kg C/Mg waste)

$k_{1,i}$ = degradation rate constant of fraction i (per year)

t = time elapsed since waste depositing (year)

10.4.2.4 European Pollutant Emission Registrar: France Model

In this model, two methods are used to quantify methane emissions from the landfills. In the first method, methane emissions from landfill cells with landfill gas recovery system are estimated by using the following equation:

$$\alpha_t = \frac{E \times H \times M_f}{R}$$

Where

E = LFG extraction rate (m^3 LFG/hour)

H = compressor yearly hours in operation (hour/year)

M_f = fraction of methane in LFG (m^3CH_4/m^3LFG)

In the second method, annual methane production is calculated with a multiphase equation using waste categories "x" of different methane generation potential as shown below:

$$\alpha_t = \Sigma_x FE_x \times \left(\sum_{i=1}^{3} k_i \times p_i \times k_i \times e^{-k_i t}\right)$$

Where

FE_x = methane generation potential (m^3CH_4/Mg waste).

p_i = fraction of rapidly, moderately and slowly degrading waste (kg_i/kg waste).

k_i = degradation rate of fraction i (per year).

t = age of waste (year)

10.4.2.5 United States LandGEM Model

LandGEM is a model developed by the U.S. EPA based on the first-order decomposition rate equation to measure landfill gas emissions from the landfills. It uses default parameters given by the Clean Air Act and the U.S. EPA's Air Pollutant Emission Factors (AP-42) for estimating emission rates in the absence of site-specific data. The main input parameters used in the model are methane generation rate constant (k), potential methane generation capacity (L_0), and annual waste disposal rates (M_i). In addition to these model parameters, LandGEM uses input parameters like landfill characteristics (landfill open year, closure year, and waste design capacity), waste acceptance rates, and selection of gas or air pollutants. LandGEM estimates the landfill gas emission from landfills using the first-order decay equation as shown below:

$$\alpha_t = \sum_{i=1}^{n} \sum_{j=0.1}^{1} kL\left(\frac{A_i}{10}\right) e^{-kt_{i,j}}$$

Where

n = (year of the calculation) − (initial year of waste acceptance)
j = 0.1 year increments
k = methane generation rate (per year)
L = potential methane generation capacity (m^3/Mg)
A_i = mass of waste accepted in the ith year (Mg)
t_{ij} = age of the jth section of waste mass A_i accepted in the ith year

BIBLIOGRAPHY

[1] Rao MN, Suresh EMS, Natesan U. Assessing site suitability conditions for waste disposal facilities using GIS for Chennai City. 14th International conference on solid waste management. In: Proceedings; 1998.

[2] Rao MN, et al. Vermicomposting-a technological option for solid waste management. J Solid Waste Technol Manage 1997;24(2):89–93.

[3] Rao MN. An ecologically balanced system for urban waste disposal. J Conserv Recy 1984;7(2):157.

[4] Sundaresan BB, Subramanyam PVR, Bride AD. An overview of toxic and hazardous waste in india. Industry and environment special issue no.4; 1983.

[5] Comprehensive Industry Document: Chlor-Alkali Industry (Comprehensive Industry Document series (Abridged): COINDS/5/1981 – 82 –A publication from Central Board for prevention and Control of Water Pollution. New Delhi.

[6] Mantell CL. Solid wastes – origin, collection, processing and disposal, 701. New York, NY: John Wiley & Sons; 1975. p. 353–64, 763.

[7] Mercer WA. Industrial solid waste – the problems of food industry. In: Proceedings national conference, of solid waste research, american public works association research foundation. Chicago, USA. p. 51–4.

[8] Bhide AD, et al. Composing of cotton dust from textile mills. Indian J Environ Hlth 1971;13:261–75.

[9] Mc Elory CT. Emplacement disposal and utilization of colliery wastes. In: Proceedings 'waste management, control, recovery and reuse' 1974, Australian waste conference held at Niv. of New South Wales; 1975. p. 171–78.

[10] American Petroleum Institute. Manual of disposal of refinery wastes. IV Solid. 1st ed. New York, NY; 1963.

[11] Singh N, Srivastava KN, Krishnan RM, Nijhavawan BR. Scope for utilization of slag and related waste from Indian iron and steel plants. In: Proceedings sump. On 'utilization of metallurgical wastes', MNL, Jamshedpur; 1954. p. 192–210.

[12] Cement Research Institute of India Utilization of Fly ash a Cement Research Institute of India 'utilization of fly ash – a technical appraisal'; 1971.

[13] Ratter BG. Waste Management in Pb – Zinc industry. Proceedings waste management, control, recovery and reuse, Fachlehragang fur mull Abfallibeseiting an der University, Stuttgart; 1974. p. 30.

[14] Sankarana C, Bhatnagar PP. Utilization of red mud, In: Proceedings symp. on 'utilization of metallurgical wastes', NML, Jamshedpur; 1954. p. 224–48.

[15] Sen MC, Banerjee T. Production of secondary A1-a review. In: Proceedings symp. on 'utilization of metallurgical wastes', NML, Jamshedpur; 1954. p. 249–54.

[16] LaGrega MD, Buckingham PL, Evans JC. Hazardous Waste Management. New York, USA: McGraw Hill; 1994.

[17] Wise DL, Trantolo DJ, Cichon EJ, Inyang HL, Stottmeister U, editors. Remediation Engineering of contaminated soils. CRC Press; 2000.

[18] Sultana R. E-waste: EPTRI, <http://whoindia.org/LinkFiles/Chemical_Safety_Report_on_Inventorization_of_e-waste_in_two_cities_in_andhra_pradesh_and_karnataka_.pdf>.

[19] Sultana R., EPTRI, Bio-Medical Waste Management (BMW) for Doctors, <http://www.whoindia.org/LinkFiles/Chemical_Safety_Bio-medical_waste_management_self_Learning_document_for_Doctors,_superintendents_and_Administrators.pdf>.

[20] Sultana R. EPTRI, bio-medical waste management (BMW) for nurses, <http://www.whoindia.org/LinkFiles/Chemical_Safety_Bio-_medical_waste_management_self_learning_document_for_Nurses_and_Paramedica.pdf>.
[21] Sultana R., EPTRI, bio-medical waste management (BMW) for dayas & aayas, <http://whoindia.org/LinkFiles/Chemical_Safety_Bio-_medical_waste_management_%E2%80%93_self_Learning_document_for_Dayas,_aayas_and_class_iv_employees_2009_.pdf>.
[22] http://www.c-f-c.com/supportdocs/cfcs.htm.
[23] http://www.intox.org/databank/documents/chemical/chromium/ehc61.htm.
[24] http://www.deh.gov.au/industry/chemicals/dioxins/pubs/incinfinal.pdf.
[25] http://www.hc-sc.gc.ca/english/iyh/environment/lead.html
[26] http://www.intox.org/databank/documents/chemical/lead/ukpid25.htm.
[27] http://www.intox.org/databank/documents/chemical/mercury/cie322.htm.
[28] http://www.atsdr.cdc.gov/toxfaq.html.
[29] IPCC. Climate change 2014: mitigation of climate change. Contribution of working group III to the fifth assessment report of the intergovernmental panel on climate change. IPCC. New York, USA; 2014.
[30] Kamalan H, Sabour M, Shariatmadari N. A review on available landfill gas models. J Environ Sci Technol 2011;4:79–92.
[31] Karanjekar RV, Bhatt A, Altouqui S, Jangikhatoonabad N, Durai V, Sattler ML, et al. Estimating methane emissions from landfills based on rainfall, ambient temperature, and waste composition: the CLEEN model. Waste Manage 2015;46:389–98.
[32] Scharff H, Jacobs J. Applying guidance for methane emission estimation for landfills. Waste Manage 2006;26:417–29.
[33] USEPA. Summary report: global anthropogenic non-CO2 greenhouse gas emissions: 1990–2030. Office of atmospheric programs, climate change division, USEPA. Washington, DC; 2012.
[34] Jackson ML. Soil Chemical Analysis. Parallel Press; 2005. p. 1–930.
[35] Black CA. Methods of Soil Analysis, Part 2. Madison, USA: American Society of Agronomy, Inc. Publisher; 1965 (Reference Book)
[36] Chilton T, Burnley S, Nesaratnam S. A life cycle assessment of the closed loop recycling and thermal recovery of post consumer PET. Res Conserv Recy 2010;54 (12):1241–9.
[37] Ritter U, et al. Radiation modification of polyvinyl chloride nanocomposites with multi-walled carbon nano tubes. MATERIALWORSENSCHAFT UND WERKSTOFFTECHNIK 2010;41(8):675–81.

INDEX

Note: Page numbers followed by "*f*" and "*t*" refer to figures and tables, respectively.

C

D

E

H

I

N

Q

R

FSC
www.fsc.org
MIX
Papier | Fördert
gute Waldnutzung
FSC® C083411

Kolibri 360 GmbH
Ferdinand-Jühlke-Straße 7
99095 Erfurt, Deutschland
produktsicherheit@kolibri360.de